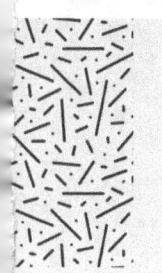

THE BUSINESS OF GOAT HERDING

SPECIALTY DAIRY,
NATURAL HERD HEALTH
+ MORE

ANNIE WARMKE
& CARIE STARR

Published by Blue Rock Station LLC
ISBN: 978-0-9791611-7-9
Text & Illustrations © 2018 Blue Rock Station LLC
Photos by Cat Harrier
Graphics by Matt Moore, Air-Loom

Contact us at:
Blue Rock Station
1190 Virginia Ridge Road
Philo, Ohio 43771 USA
Telephone: +1 (740) 674- 4300, Email: annie@bluerockstation.com
www.bluerockstation.com

Table of Contents

INTRODUCTION	1
WOMAN'S STORY – SASHA SIGETIC	3
SETTING THE COURSE	5
GOAT DO'S & DON'TS	8
BASICS OF HUMANE CARE	12
BASIC PLANNING AND AWARENESS FOR HERD MANAGERS	13
PLANNING FOR ANIMAL DEATH – OR INJURY AND DISEASE	14
STANDARDS AND CERTIFICATIONS	15
RECOGNIZING A HEALTHY GOAT	17
SPECIALTY DAIRY FARMING RESOURCES POSTER	20
WOMAN'S STORY: LESLIE SCHALLER	21
COMMUNITY SUPPORTED AGRICULTURE (CSA)	21
SELLING RAW MILK	24
ORGANIC STANDARDS	26
RAW MILK - THE REAL FACTS	27
LIST OF SPECIALTY DAIRY PRODUCTS	29
I WANT TO MAKE CHEESE POSTER!	34
CHEESE MAKING THE REALLY OLD-FASHIONED WAY	35
I WANT TO BE LICENSED!	38
PREPARATION OF HIDES FOR TANNING	39
WOMAN'S STORY: MICHELLE GORMAN	43
CALCULATING COSTS: LEARNING HOW TO SET PRICES	45
PRICING STRATEGIES	47
WOMAN'S STORY: ABBE TURNER	51
THE ART OF SIMPLE BRANDING AND MARKETING	54
CROWDFUNDING	58
SOCIAL MEDIA	63
GOAT GUIDE TO LIFE POSTER	72
WOMAN'S STORY: ANNIE WARMKE	73
ANIMAL HEALTH ISSUES	74
BASIC REQUIREMENTS – MINERALS	77
PRO-BIOTICS, NOT ANTI-BIOTICS	82
FIRST AID: THE GREAT EIGHT	83
ESTIMATING WEIGHT	87
HERBAL REMEDIES: WHAT THEY DO	88
PARASITE CONTROL	91
DEWORMER CHART FOR GOATS POSTER	96
HERBAL WORMER RECIPE	97
PASTURE MANAGEMENT	101
FENCING FOR GOATS	105
EXAMPLE OF A SOIL SAMPLE REPORT	109
THE MEDICINE CHEST	110
WOMAN'S STORY: CARIE STARR	113
WOMEN GROW OHIO - WOMEN MENTORING WOMEN	115
WOMAN'S STORY: CELESTE TAYLOR	117
SPECIALTY DAIRY FARMING RESOURCES FOR GOATS POSTER	120
PODCASTS AND WEBINARS	122
RESOURCES	123
CONTRIBUTOR BIOS	128
ABOUT THE AUTHORS	133

Dedications

"To Grandma, the original wild woman who paved my way."
- **Carie Starr**

"To Minkster, for giving me the title and so much more."
- **Annie Warmke**

INTRODUCTION

Often the business of dairy livestock requires the farmer to be the head of the herd. The herder must provide the animal with the tools they need to support their natural system by making sure that they receive the right mix of food (and vitamins/minerals), clean water and a healthy environment.

Challenges exist for many reasons, including lack of access to veterinary care, business knowledge, or where to begin to make a living. Ruminants (cows, sheep, goats) and humans can create a winning combination if the animal has what it needs to live a long healthy productive life. The humans receive in turn food, income, manure to build soil health, and healthy offspring... more importantly, the human soon understands how to work within the animal world, and the world of health regulations and business rules.

"Goats are amazing creatures. They are the most widely kept animal in the world."

Goats are amazing creatures. They are the most widely kept animal in the world. The interest in goat herding is growing daily with the increase in the immigrant community and its demand for goat meat, hides and young animals for butcher. Even the bones and manure have uses and present possibilities for revenue. But making a living from goat herding and specialty dairy is much more than just knowing that a certain demand for these products exists. The question this book seeks to answer, is where does one begin in a business that uses ruminant livestock to make a living?

Since women tend to be the goat herders and are entering the agriculture field at an increasing rate, it is important to recognize that research demonstrates that women do things differently. The growing trend, according to Dr. Rachel Terman, is that women farmers and homesteaders have low or no debt, work together more, farm on a smaller scale overall, and raise more on less land than do "traditional" farmers. This book is for those women.

Traditionally women have learned from each other through word of mouth. This book shares some stories of how women got started and how they make a living. Passing on the inner workings of farming has become something that lately seems to be held close to the chest, and topics, such as how-to-make a living, have been kept out of the discussion, according to Becky Rondy of Green Edge Gardens.

This book touches briefly on many aspects of the world of goats. Don't read it like a novel, but rather skip around and use it as a reference tool when and how each topic arises in the goat raising journey.

> "Traditionally women have learned from each other through word of mouth."

The goal is not to teach everything there is to know, making the basics so complicated that getting started seems daunting. A little knowledge, a little experience and some time to digest and make sense of information goes a long way towards succeeding in business.

Many thanks to NCR SARE Research and Education, Rural Action, ACEnet, and the people who work in these organizations. Without them and their vision it would be far more challenging to be in the business of agriculture.

Thank you Sasha Sigetic (Black Locust Livestock & Herbal) for all of your inspiration. Lots of women agriculture producers went beyond their comfort level to be interviewed: Michelle Gorman (Integration Acres), Abbe Turner (Lucky Penny Farm), Morgan Phelps (Highland Haven Farm), Becky Rondy (Green Edge Gardens), and Celeste Taylor. Special thanks to Leslie Schaller and Tom Redfern who saw the value of bringing women farm producers to the table and that has led to a broad discussion on many levels. We need more people like you in our world.

WOMAN'S STORY – SASHA SIGETIC
Sasha Sigetic, Owner of Black Locust Livestock and Herbal

I am the farmer and owner of Black Locust Livestock and Herbal, where I have a raw milk herd share with Guernsey goats, as well as grow herbs and create herbal products for animals and humans.

In Ohio, farmers cannot legally sell raw milk directly to consumers. Raw milk can only be sold through a herd share as a way for people to collectively own an animal and consume the raw milk from their own animals. For urban and suburban people, herd share becomes a way to pay a farmer to board and milk an animal that they collectively own. Through this legal process, they have access to raw milk.

My interest in goats began when I lived in Austin, TX, where I took permaculture design certification courses. I already had a substantial background in nutrition and natural health plus I also really liked goat milk.

From these courses the idea of keeping goats (even though I had never had one before) was born. Through the courses I realized there is a wonderful way of keeping a bunch of species in a relatively small space using closed-loop cycles. For example, a farmer can use the manure from the goats in the orchards as well as run the goats through the orchard so they can browse and clean up. In this way the goats provide services for the farmer in addition to providing beautiful, delicious milk.

When I moved to southeast Ohio, I started working with Integration Acres, milking goats for their creamery business. Eventually I began keeping my own goats.

My model for milking includes drying off the milk goats for the winter.

Some people milk their goats year round, which requires the goat herder split the herd in half and breed half every other year. But because I farm alone, I find that I need to give myself a little bit of a break during the winter.

Farming is generally not a single-person job, so there's a lot of value in having neighbors and a community of people around to assist in handling tricky situations. If, for example, a goat gets out of a fenced enclosure, having an animal person nearby is really helpful.

> "If you have livestock, you will have dead stock. Death is part of the cycle of life."

If you're interested in starting a goat herd, I believe it is best to read a lot of good books and build a strong fence. Invest in the goats' infrastructure, because goats are smart. They will test fences, gates and your patience. The trade-off is that they have awesome personalities. They enjoy their farmer, a lot like pets enjoy their owners.

The hardest part of raising goats is to learn to be okay with death. If you have livestock, you will have dead stock. It is vital to understand

that death is a part of the cycle of life-- accidents, sickness, genetic defects happen.

Nature isn't kind, but it sure is honest.

SETTING THE COURSE

"I am not going to be like Christopher Columbus. He didn't know where he was going, he didn't know where he was when he got there, and he didn't know where he had been when he returned home. And along the way, he did a lot of damage." Annie Warmke

→ What will I need to sacrifice (or did I already make the sacrifices)?
→ What excuses have I used for not setting or completing my goals in the past?
→ What five people can I call on for help and support?
→ What special skills do I possess to reach my goals? (ie: knowledge of horsemanship, gardening, people person, livestock management, cooking, knowledge about history of the area, etc.)
→ What skills will I need to develop to reach my goals? How will I accomplish this?
→ What motivates me?
→ What will be the rewards if I work towards this goal?
→ Who will my goals benefit? (Me? My family? My work? My future? My neighborhood? My community?) In what way?
→ Who are my allies?
→ Who are my adversaries?

"Raising goats is a business, and every business needs a plan. Ask yourself these questions."

Rate family members or other people who may be connected to your business:
→ Are they friendly, confident, tolerant of strangers/guests, supportive?
→ What skills and resources do they bring to my business ideas?

Accountant _____

Attorney _____

Lender/Financier _____

Consultant _____

Veterinarian _____

Neighbors: Will they be supportive?

Sheriff and deputies: Will they be supportive, or suspicious of outsiders?

Game Warden

Government officials, inspectors, regulators (local, state, federal)

Economic Development Staff (local, county, regional, state)

Tourism agencies

Local businesses

Other farmers in a similar business

> *"There are many people that will affect your business. What skills and resources do they bring?"*

→ What are my greatest fears?
→ What are the risks involved?
→ What are the obstacles?
→ *What resources are available?*

Physical Resources/Land Resources — list what you have (high speed Internet, TV/radio stations, local newspapers, deeded, leased private property, state & federal allotments, range land, woodland, hay meadows, cropland, riparian/wetlands, acreage, location proximity, elevations, topography, location of feeding grounds, parking areas, etc.)

Climate — consider how weather patterns, temperature, etc. will affect what types of activities you might provide: Temperature variations;

length of growing season; precipitation (monthly distribution): or snow (ground cover & accumulation depth).

Developments & Improvements (buildings, fences, corrals and working facilities, equipment, roads and trails, etc.).

Other Attractions — list those things that might enhance your operation or product to a customer, such as wildlife streams, ponds, fishing, livestock, proximity to natural or man-made points of interest, organic pastures, historic buildings, etc.

If agri-tourism is in the plan, are there other attractions in the area that visitors can visit?

Potential Hazards — (Farm equipment, storage, areas that need physical improvements, wetlands).

> *"What other resources are available that will help this business to succeed and thrive?"*

- → How long will it take to make a living from this business?
- → How long can I sustain myself financially until I can make a living with this business?
- → What else can I do to make money while I am building the business?
- → What are the resources I've listed that could generate revenue? How much at the beginning of the business? How much over longer periods of time?
- → How persistent will I need to be to reach my goal? (work long hours, borrow money, building repairs or construction, hire employees, gain the trust of the community or neighbors).

Goat Do's & Don'ts

Goats have very specific needs. If the farmer gets those needs wrong, the result often leads to ill health, and/or death. Whether the goat herder is new to goats, or been at it a long time, these guideline basics have the potential to provide a broad base of tools that can keep goats healthy, and happy.

> *"These guidelines have the potential to provide a broad base of tools that can keep goats healthy and happy."*

The Basics for all Animal Welfare

→ Access to wholesome and nutritious feed.
→ Appropriate environmental design.
→ Caring and responsible planning and management.
→ Skilled, knowledgeable, and conscientious animal care.
→ Considerate handling, transport, and slaughter.

What's Normal in the Goat World?

Goats are prey animals, which means they know they are being hunted by predators - so they are naturally agile and flighty, ready to scatter and/or face the enemy. Stress, such as being chased by a predator or poor living conditions (including: not enough space for resting, dirty housing conditions and bad feed) lowers their conception rates and reduces both immune and rumen functions.

The wisest doe in the herd is called the "Queen". She selects herself to be the leader (she may have to fight for the title) and she is the one to move first when the herd is threatened or simply changes grazing location.

The lead goat eats first, decides what will be eaten in the field, and sleeps where she wants. Goats are foragers – they eat leaves off bushes, trees, or weedy plants. They need a mixture of plants in their diet. Regular pasture rotation is necessary to avoid parasite overload in the goat gut.

Goats gather together while they are ruminating, usually resting in their favorite spots to chew their cud, and watch for any problems, or bathe in the sun for some Vitamin D. When they are foraging, they tend to spread out.

Kids seek bodily contact with dams, and dams nuzzle their kids in various ways as a sign of affection. Dams respond to their young's calls just after birth by letting down their milk so nursing can take place. They can also be affectionate with other goats, adult or kid, by rubbing their heads and necks (sometimes this is also done as an act of domination).

Goat skin is sensitive, even though covered with hair. Their nose, mouth, lips, ears and foot pads easily feel touch, and generally they like being massaged, petted and brushed.

> *"Goats know who they like within the herd and among humans. They can learn patterns and use of repetition..."*

Goats will not eat dirty food or water unless they are starving. They use their sense of smell to identify other goats, detect estrus (bucks), water location, plus differences between pastures and feeds.

HERD PSYCHOLOGY

Goats know who they like within the herd and among humans. They can learn behavior patterns through repetition, so they can do things like wait to be tied up at their lead.

A female is always the leader of the herd and the Queen's kids are high in the pecking order.

When a new goat, or even a new human, comes into their space, the dominant goats will challenge the newcomer in some form or fashion.

Goats within the hierarchy often posture for a higher place in the herd. This usually occurs when a goat leaves the herd, or after kidding season.

It's easy for goats to learn their names and acknowledge they are being called. They will often respond with a grunt to acknowledge they heard their name called. They vocalize frustration, pain, stress, separation from kids, food being made available and will warn other herd members of an intruder.

> "Goats within the herd hierarchy often posture for a higher place in the herd. This usually occurs when a goat leaves or after kidding season."

They naturally move into larger areas when available because they don't like tight spaces. Goats prefer to move up an incline, rather than down.

They don't like to be startled and can drop dead from a loud noise such as a dog barking at them. Goats move toward light, not dark or uneven lighting.

They have good long-term memory, especially if something startled or scared them in the past.

Studies in Australia show goats can even be trained to use a sand box of sorts to pee and poop.

BASICS OF HOUSING/FENCING

The space where a goat lives, eats and wanders needs to be free of objects and situations that can cause injuries or bruising to goats.

Of course they are capable of surviving occasional bumps and bruises, but these should generally come from other goat interactions. The living space should not allow the goat to come into contact with toxic fumes or surfaces treated with toxic paint, wood preservatives, etc.

Goats should not have access to anything electrical, and the electrical wiring should be protected against rodents.

Outdoor and indoor lying areas should be easily accessible, and have dry bedding, plus be of sufficient size for all goats to lie down. Total floor space should not be less than 1.5 times (per goat) their minimum lying area.

Housing for kids should include dry bedding and have effective ventilation. When the temperature falls low enough to cause the animals to shiver uncontrollably, supplementary heating should be made available, especially for very young animals.

When goats are being reared on pasture, they must all have adequate shelter, either natural or man-made, to protect them from wind, rain and extreme heat or cold. In winter, additional shelter or windbreaks are absolutely necessary.

> *"Pen shape and space allowance need to provide sufficient freedom of movement to permit exercise."*

Pen shape and space allowance needs to provide sufficient freedom of movement to permit exercise. Minimum lying space allowances for typical dairy breeds are as follows:

- Adult does – 18 square feet
- Bucks – 30 to 40 square feet
- Kids – 8 to 10 square feet

Bucks need to be housed with other goats or at least within sight and sound of herd mates. Otherwise they can become very stressed, as they do not feel safe outside of the herd's protection.

Electric fences need to be designed, installed, used, and maintained so that contact with them does not cause more than momentary discomfort to the goats. Electric mesh fencing is very dangerous for horned goats. Fences must be designed to withstand climbing and prevent injury to goats.

Basics of Humane Care

In order for goats to be healthy and able to fight off disease and parasites, they need to be fed a daily ration of feed that takes into consideration their dietary, vitamin and mineral requirements.

> *"It is important to avoid sudden changes in type and quantity of feed... the rumen cannot adjust..."*

These requirements are unique to goats. Goats must be provided with feed or forage containing adequate, suitable fiber to allow the rumens to operate properly. It is important to avoid sudden changes in type and quantity of feed, unless directed by a veterinarian. Sudden changes in feed prevent the rumen from digesting the food properly.

When climatic and geographic conditions are right, goats should have voluntary access to pasture or an outdoor exercise area.

Stored feeds, such as good quality hay and silage, need to be protected from rodents and other animals. Anything that poops on the hay will make it so the goat will not eat it.

Kids should not be weaned before six weeks of age and should have access to dry feed (hay) from two weeks of age on to encourage proper rumen development.

BASIC PLANNING AND AWARENESS FOR HERD MANAGERS

Many people get into goat herding without a basic knowledge of the animal's needs, and their own role in keeping things going smoothly for both the goat and the herder. The goat herder needs to have an understanding of how to deal with many situations that will arise sooner or later.

There are many good books, YouTube videos, podcasts, and webinars (see the resource section for suggestions) that can provide information and reliable knowledge on the basics. Or attend a goat school for hands-on experience.

The basics include:

→ Kidding (including how to give colostrum and how to avoid the problems of goat bad mothering, care of the newborn kid)
→ Injections
→ Drenching
→ Dehorning
→ Castration
→ Shearing
→ Milking procedures
→ Hoof trimming and maintenance
→ Euthanasia
→ Death
→ Slaughtering
→ Emergency preparedness
→ Recognize signs of normal behavior, abnormal behavior, pain and fear
→ Recognize signs of common diseases, understand their prevention and control, and know when to seek veterinary help
→ Have a basic knowledge of what constitutes proper nutrition in goats
→ Have knowledge of body condition scoring
→ Understand functional anatomy of the normal foot, its care and treatment

"Recognize signs of normal behavior, abnormal behavior, pain and fear."

PLANNING FOR ANIMAL DEATH – OR INJURY AND DISEASE

Provisions need to be in place for dealing with the segregation and care of sick and injured animals when needed to prevent further injury or spread of a contagious condition.

Any goats suffering from a contagious illness or susceptible to further injury should be:

> *"Provisions need to be in place for dealing with the segregation and care of sick and injured animals..."*

→ Segregated
→ Treated without delay
→ Able to benefit from veterinary advice when needed; or, when necessary, humanely euthanized.
→ A plan for disposing of urine and dung from hospital pens housing sick and injured animals to avoid spreading infection to other stock.

STANDARDS AND CERTIFICATIONS

Why certify your business? Well... it's good for business.

Consumers are increasingly interested in, and willing to pay more for, animal products from farms raising animals with more humane animal husbandry practices. Retailers, distributors, food service companies and restaurants are increasingly seeking to buy welfare-certified animal products.

Meanwhile, major media outlets and consumer organizations are addressing farm animal welfare and helping consumers navigate food labels. These trends present farmers and ranchers with the opportunity to benefit by providing evidence of more humane animal care.

> *"Studies show that 84% of consumers view better living conditions for animals as very important..."*

Studies show that 84% of consumers view better living conditions for animals as very important in their choosing products, while 74% say they are paying more attention to labels regarding animal living conditions than they were five years ago (*2015 Consumer Reports*).

Certifications information is available at:
National Agricultural Library (NAL). USDA. Animal welfare information center (n.d.) Retrieved from https://www.nal.usda.gov/awic/certification-programs

The focus of certification programs is to provide information, products, services, and activities that promote the humane care.

Animal Welfare Approved (AWA)
AWA audits, certifies and supports independent family farmers raising their animals according to the highest animal welfare standards, outdoors on pasture or range. AWA is a program of A Greener World, a non-profit organization dedicated to promoting sustainable practices for farmers and ranchers.

Humane Farm Animal Care (HFAC) is the leading non-profit certification organization dedicated to improving the lives of farm animals in food production from birth through slaughter by driving consumer demand for kinder and more responsible farm animal practices.

SUGGESTED RESOURCES

- → National Agricultural Library (NAL). USDA. Animal welfare act (n.d.). Retrieved from https://www.nal.usda.gov/awic/animal-welfare-act
- → North Central SARE. (n.d.). North Central sustainable agriculture research and education program- Grants and Education. Retrieved from https://www.northcentralsare.org
- → Food Animal Concern Trust FACT. (n.d.). Retrieved from https://foodanimalconcernstrust.org
- → Michigan State University Extension. MSU (2014, March 17). Animal welfare for youth. Retrieved from http://www.canr.msu.edu/news/animal_welfare_for_youth_part_1_what_is_animal_welfare

Recognizing a Healthy Goat

When Eleonore Rigby and her two tiny kids arrived at Blue Rock Station (BRS), it would have been hard to imagine the adventures and challenges to come, partly because Eleonore came from a "rough" environment and had never been outside of a barn.

Over the course of her first years, she had a dry skin condition every winter (from lack of sunshine), bloat from getting into the feed and eating all she wanted (twice), and skin parasites.

With the right combination of nutrients, she gained ground each year (her natural immune system growing stronger and stronger). But because she started out in such a poor environment prior to arriving at BRS, she faced a constant battle to stay in balance.

> "Despite the fact that goats are one of the oldest and most widespread domesticated animal, there is surprisingly little research into their health."

Any change (in routine, amount of light, nutrition) created minor health issues that had to be immediately addressed. Eleonore was a living example that a healthy environment from DAY-ONE is vital to life-long health.

Despite the fact that goats are one of the oldest and most widespread domesticated animals, there is surprisingly little research into their health. Books about farm animals generally ignore goats - leaving it to individual goat herders to learn and recognize what is "normal" within their herd.

When a goat is ill, there are basic signs or signals.

It is a good idea to keep a notebook or journal to record some of the behaviors and important events that occur in each goat's life. These notes can be useful over time, since memory's fade.

> *"Understanding the basics of what makes up a healthy goat may also be helpful when considering buying a goat."*

Ashley (another early goat at Blue Rock Station's herd), for example, had a habit of pointing her nose into the air and waving her head about when she was nervous (or in heat). For her, this was normal. In another goat it might be a sign that something is seriously wrong.

Understanding the basics of what makes up a healthy goat may also be helpful when considering buying a goat to add to the herd.

A few items to note include:

- **Attitude** - Alert and inquisitive
- **Breathing** - Even
- **Coat** - Clean, glossy and no lumps
- **Eyes** - Bright and not running
- **Feed** - Likes to eat (it is a goat after all). Chews cud after eating
- **Gait** - Steady, with no limping
- **Nose** - Cool and dry
- **Poop** - Firm and pelleted (looks like plump raisins) with no signs of parasites
- **Reproductive Organs** – Everything is present and accounted for
- **Urine** - Light brown and clear

- → **Udder** – Two teats that hang evenly (no extra holes)
- → **Weight** - Normal weight for the breed
- → **Temperature** - A goat's normal body temperature is usually in the range between 101.5° and 103.5° F. Young goats usually have higher temperatures than adult goats. To determine a goat's normal temperature, it is a good idea to record the body temperatures of goats in the herd that seem healthy and are about the same size. The temperature is taken with a rectal thermometer, inserted for approximately three minutes. If a goat has a temperature above or below the normal range, this is usually an indication that there is something wrong. A higher than normal temperature probably means there is some sort of infection. *NOTE: Some experts say that a temperature of 106° F for an extended time with bucks can cause sterilization.*

SUGGESTED RESOURCES

North Caroline State University. Basic Meat Goat Facts (2015, September 21). Retrieved from https://content.ces.ncsu.edu/basic-meat-goat-facts

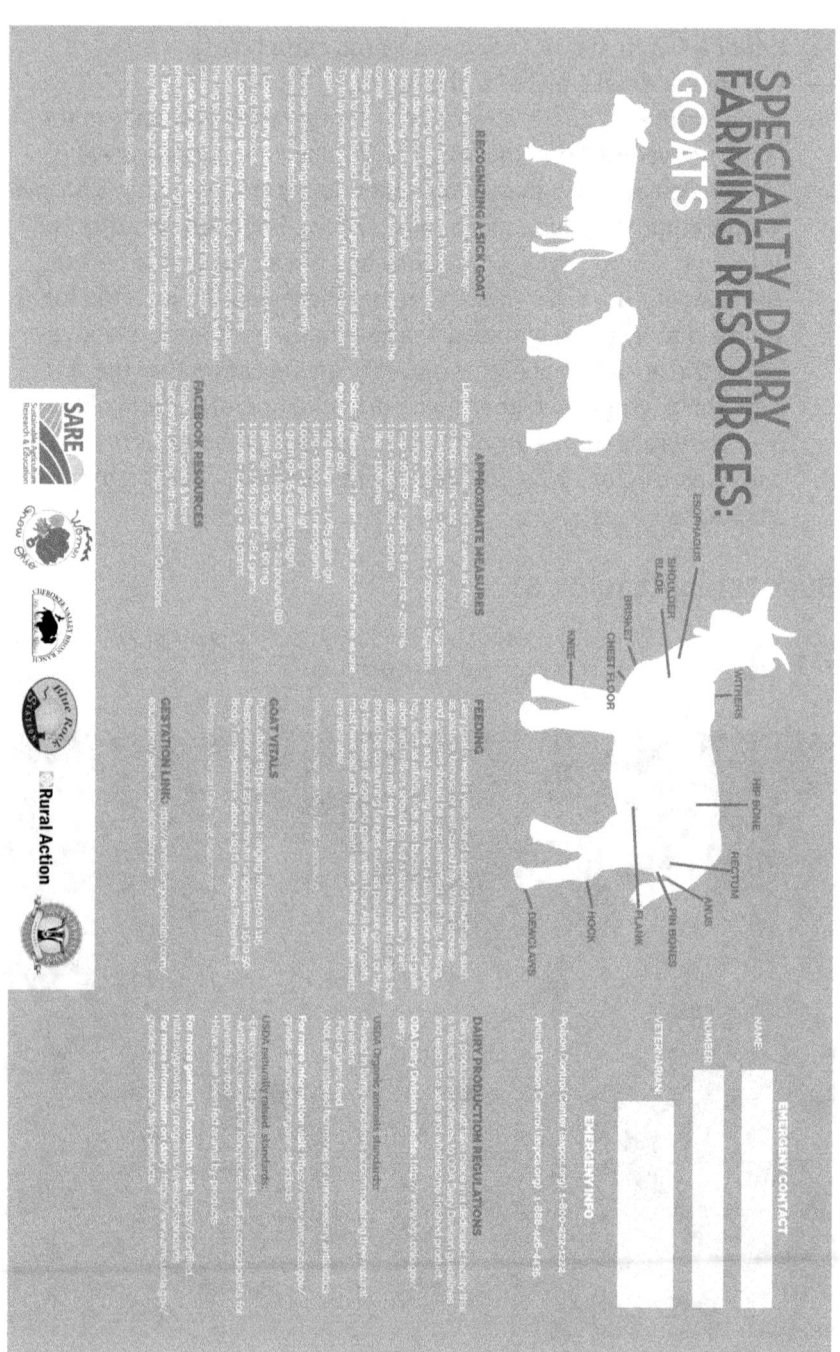

Woman's Story: Leslie Schaller

Since 1985, the year ACEnet began, I have been involved in the local food economy movement. I jokingly call myself a "failed farmer" although I still garden. I have become well-known as a local "food enabler".

For nearly 50 years I have lived in the southeast Ohio region and I remain inspired by the role of women and their regional heritage of learning from one another. In the late 70's and 80's it was really the women who were more available to share their knowledge.

As someone who appreciates goats and loves their many products, I see a tremendous opportunity for informing farmers, both male and female, on how to do season creation (adding a winter growing season), which is a model that lends itself well to artisanal dairy.

My passion for the local food economy has been inspired by watching the grit and "stick-to-it-ness" of so many women farmer role models helping other women learn from each other.

Regulations/Standards
Community Supported Agriculture (CSA)

The CSA model is a system that connects food producers and consumers by allowing the consumer to purchase a subscription for high quality, often organic foods from a farm or group of farms. It allows the producer and consumer to share the risks of farming.

In return for signing on to a harvest, subscribers receive either a weekly or bi-weekly box of produce or other farm goods such as herbs, eggs, meat or milk.

Typically, farmers try to cultivate a relationship with subscribers by sending weekly newsletters of what is happening on the farm, inviting them to pick their own produce, or holding an open-farm event. Some CSAs provide for contributions of labor in lieu of a portion of subscription costs.

> *"A CSA develops a consumer group that is willing to fund a whole season's budget."*

The basic premise of a CSA includes developing a consumer group that is willing to fund a whole season's budget in order to purchase quality foods. The system has many variations on how the farm budget is supported by the consumers and how the producers then deliver the foods. CSA theory supports the idea that the more a farm embraces whole-farm, whole-budget support, the more it can focus on quality and reduce the risk of food waste.

CSA farms share three common characteristics:
→ an emphasis on community and/or local produce,
→ share or subscriptions sold prior to the farm growing/producing season,
→ weekly deliveries or pickups to members/subscribers.

The functioning of a CSA also relies on four practical arrangements:
→ for farmers to know the needs of a community,
→ for consumers to have the opportunity to express to farmers what their needs and financial limitations are,

→ for commitments between farmers and consumers to be consciously established,
→ for farmers' needs to be recognized.

From this base, four main types of CSAs have been developed:
→ Farmer managed: A farmer sets up and maintains a CSA, recruits subscribers, and controls management of the CSA.
→ Shareholder/subscriber: Local residents set up a CSA and hire a farmer to grow crops, and shareholders/subscribers control most management.
→ Farmer cooperative: Multiple farmers develop a CSA program.
→ Farmer-shareholder cooperative: Farmers and local residents set up and cooperatively manage a CSA.

CSAs can have a core group of members that help to make decisions about and run the CSA, including: marketing, distribution, administrative, and community organization functions.

> "CSAs with a core group of members operate more successfully as a farmer-shareholder cooperative."

CSAs with a core group of members operate more successfully as a farmer-shareholder cooperative and CSAs without a group of core members rely much more on subscriptions and run most prominently as shareholder/subscriber CSAs.

How it Works

Shares are distributed in several different ways, but usually distributed weekly. Most CSAs allow share pickup at the farm. Shares are also distributed through regional dropoff locations, direct home or office drop-off points, farmers' markets, and community center/church dropoff centers.

CSAs market their farms and shares in different ways. They often use local farmers' markets, restaurants, on-farm retail shops, wholesale to natural food stores, and wholesale to local groceries, in addition to their CSAs to market shares.

One problem that CSAs encounter is over-production. CSAs often sell their produce and products in ways other than shares to distribute the extra goods. CSA farms also sell their products at local farmers' markets. Excess products are sometimes given to foods banks.

> *"If a goat herder is thinking of selling raw milk, then the place to start is the Farm to Consumer Legal Defense Fund."*

Challenges for Farmers

Many CSA farmers can capitalize on a closer relationship between customers and their food, since some customers will pay more if they know where it is coming from, who is involved, and have special access to the farmer and the places the products are produced.

SELLING RAW MILK

In most states in the U.S. it is illegal to sell raw milk. In some states, like Ohio, it is possible to sell raw milk through the herd share model. If a goat herder is thinking about selling raw milk, then the place to start is the Farm to Consumer Legal Defense Fund (FCLDF). They can provide information on local laws, cottage industry regulations, and a legally binding herd contract.

Membership in FCLDF is open to family farms throughout the United States that engage in non-toxic and/or grass-based farming and do not engage in practices that endanger the environment or harm the health of others.

FCLDF also provides members with: potential legal representation to defend distribution of raw milk and other farm products directly to the consumer; participation in direct commerce to make lobbying and advocacy available in the event of government interference; advice on federal, state and local agriculture and health department regulations affecting direct distribution of farm products; and access to information tailored to specific needs and/or location.

HERD SHARE

A herd share, (farm share, cow share, goat share, etc.) is where people become owners of a portion of one or more dairy animals by buying shares of a milking animal or herd. In exchange, the farmer cares for the animals and milks them. As owners, the shareholders are entitled to the milk (or other related products) from their animals.

With herd share the farmer is not selling milk to customers; instead the "customer" is receiving milk from their own dairy animal that has been purchased through a contract.

The process is similar to how shares of race horses are sold. Someone wants to own a race horse, but it's quite expensive, or they don't have a stable for housing the animal.

They invest in the horse by purchasing a portion of it, and then pay monthly fees for board, train and keep of the animal. Any winnings are then shared amongst the owners, based on the portion of the animal owned.

"A herd share is where people become owners of a portion of one or more dairy animals... and are entitled to the milk from their animals."

With a livestock herd share, the farmer determines the exact cost of keeping the dairy animal per day or month. The consumer (new owner of one or more animals) purchases shares based on this cost, and the farmer then maintains/boards the animals in exchange. There are also monthly maintenance fees typically charged for the boarding of the animal.

The new owner of that portion of the herd (or single animal) receives a predetermined portion of the milk and/or other related products.

SUGGESTED RESOURCES

- → Agricultural Marketing Service. USDA. (n.d.). Local Food Directories: Community Supported Agriculture (CSA) Directory. Retrieved from https://www.ams.usda.gov/local-food-directories/csas
- → Farm-to-Consumer Legal Defense Fund - Defending the rights and broadening the freedoms of family farms and protecting consumer access to raw milk and nutrient-dense foods. (n.d.). Retrieved from https://www.farmtoconsumer.org
- → Farm-to-Consumer Legal Defense Fund. Top 10 Herd Share Questions Answered (2017, October 1). Retrieved from https://www.farmtoconsumer.org/blog/2017/10/01/top-10-herd-share-questions-answered

ORGANIC STANDARDS

> *"Organic dairy producers must follow production standards set by the U.S. Department of Agriculture."*

Organic dairy producers must follow production standards set by the U.S. Department of Agriculture. USDA Organic animals standards:

→ Raised in living conditions accommodating their natural behaviors

→ Fed organic feed

→ Not administered hormones or unnecessary antibiotics

For more information on how to become organically certified, visit: OEFFA (oeffa.org). OEFFA certification currently accepts new applications from operations in Ohio, Michigan, Kentucky, Indiana, West Virginia, Pennsylvania, Iowa, Illinois, Missouri, and Virginia.

SUGGESTED RESOURCES

- → Agricultural Marketing Service. Organic Standards (n.d.). Retrieved from https://www.ams.usda.gov/grades-standards/organic-standards
- → Ohio Ecological Food and Farm Association OEFFA. (n.d.). Retrieved from https://www.oeffa.org

RAW MILK - THE REAL FACTS
(Reprinted from Weston A. Price Foundation)

Raw milk is a living food...nutritious and easy to digest.

For centuries traditional cultures have eaten raw and cultured raw milk products from cow, sheep, goat, camel, yak, water buffalo and reindeer milk. Weston A. Price studied many of these traditional cultures and found that people were vibrantly healthy and had perfect teeth.

In Dr. Price's book *Nutrition and Physical Degeneration*, photos clearly illustrate the physical degeneration that occurs when people abandon nourishing traditional diets in favor of modern, process convenience foods. Galen, Hippocrates, Pliny, Varro, Marcellus Empiricus, Baccius and Athimus, leading physicians in their day, all recommended raw milk for various ailments.

> "Leading physicians in their day all recommended raw milk for various ailments."

During the 1920s, Dr. J. E. Crewe of the Mayo Foundation used a diet of raw milk to cure TB, edema, heart failure, high blood pressure, prostate disease, urinary tract infections, diabetes, kidney disease, chronic fatigue and obesity. Today, in Germany, successful raw milk therapy is provided in many hospitals.

Real Milk comes from real cows that eat real feed.

Real feed for cows is green grass in spring, summer and fall; stored dry hay, silage, hay and root vegetables in winter. It is not soy meal, cottonseed meal or other commercial feeds, nor is it bakery waste, chicken manure or citrus peel cake, laced with pesticides.

Vital nutrients like Vitamins A and D, and Price's "Activator X" (a fat-soluble catalyst that promotes optimum mineral assimilation, now believed to be Vitamin K2) are greatest in milk from cows eating green grass, especially rapidly growing green grass in the spring and fall.

Vitamins A and D are greatly diminished, and Activator X disappears when milk cows are fed commercial feed. Soy meal has the wrong protein profile for the dairy cow, resulting in a short burst of high milk production followed by premature death.

- → Real Milk is not homogenized.
- → Real Milk contains no additives.
- → Real Milk contains butterfat, and lots of it.

Suggested Resources

- → *https://www.westonaprice.org/wp-content/uploads/RealMilkBrochure.pdf*

LIST OF SPECIALTY DAIRY PRODUCTS

Dairy products are made from the milk of cows, goats, sheep, water buffalo, moose, horses, donkeys, reindeer, and camels plus they include hundreds of types of cheeses, cultured milk, fermented milk, ice cream, cultured dairy, yogurt, kefir, whey protein, butter and frozen desserts. The use of various types of milk has been around for thousands of years, starting with carrying milk from place to place in a goat or camel stomach.

With the growth of immigrant communities, the demand for more variety in dairy products is growing. An extensive list of dairy products can be found at https://en.wikipedia.org/wiki/List_of_cheeses or https://en.wikipedia.org/wiki/List_of_dairy_products

Aarts	Dried fermented milk often mixed with various measures of sugar, salt or oil. Eaten as a snack or reconstituted as a warm beverage.
Acidophiline	A drinkable yogurt, with Lactobacillus acidophilus as the starter culture.
Anthotyros or Anthotyro	Anthotyros is Greek traditional cheese, which is prepared by adding goat-sheep's milk or milk cream to sheep or goat cheese.
Blaand	A fermented milk product made from whey. It is similar in alcohol content to wine.
Black Kashk	Prepared from yogurt, its production involves several processes.
Booza	An elastic, sticky, high level melt resistant ice cream.
Buffalo Curd	A traditional and nutritious dairy product prepared from water buffalo milk .
Buttermilk koldskål	A sweet cold beverage or soup, made with buttermilk and other ingredients.
Cacik	A Turkish dish of seasoned, diluted yogurt.
Caudle	A British thickened and sweetened alcoholic hot drink, somewhat like eggnog.
Chaas	A buttermilk preparation from India containing raw milk, cream (malai) or yogurt which is blended manually in a pot with an instrument called madhani (whipper).
Chal	Fermented camel milk, sparkling white with a sour flavor.

Chalap	A beverage made of yogurt, salt, and modernly, carbonated water.
Chass	It is similar to Lassi.
Cheese	A food derived from milk that is produced in a wide range of flavors, textures, and forms by coagulation of the milk protein casein. It comprises proteins and fat from milk, usually the milk of cows, buffalo, goats, or sheep.
Clabber	Produced by allowing unpasteurized milk to turn sour at a specific humidity and temperature. Over time, the milk thickens or curdles into a yogurt-like substance with a strong, sour flavor.
Clotted cream	A thick cream made by indirectly heating full-cream cow's milk using steam or a water bath and then leaving it in shallow pans to cool slowly. During this time, the cream content rises to the surface and forms 'clots' or 'clouts'. It forms an essential part of a cream tea.
Condensed milk	Milk from which water has been removed. It is most often found in the form of sweetened condensed milk, with sugar added.
Cottage cheese	A cheese curd product with a mild flavor. The curd is drained from the whey, but not pressed, then washed to get rid of any whey residue. The individual curds remain loose.
Cream	Composed of the higher-butterfat layer skimmed from the top of milk before homogenization. In un-homogenized milk, the fat, which is less dense, will eventually rise to the top.
Cream cheese	A soft, mild-tasting cheese with a high fat content. Traditionally, it is made from unskimmed milk enriched with additional cream. Stabilizers such as carob bean gum and carrageenan are added.
Crème anglaise	A light pouring custard used as a dessert cream or sauce. It is a mix of sugar, egg yolks and hot milk, often flavored with vanilla.
Crème fraîche	A soured cream containing 30–45% butterfat and having a pH of around 4.5. It is soured with bacterial culture.
Cuajada	A (milk curd) cheese product. Traditionally it is made from ewe's milk, but now it is more often made industrially from cow's milk.
Curd	Curd is obtained by curdling (coagulating) milk with rennet or an edible acidic substance such as lemon juice or vinegar, and then draining off the liquid portion.
Custard	A variety of culinary preparations based on a cooked mixture of milk or cream and egg yolk. Depending on how much egg or thickener is used, custard may vary in consistency from a thin pouring sauce to a thick pastry cream used to fill éclairs.
Dadiah	A traditional fermented milk made by pouring fresh raw unheated buffalo milk into a bamboo tube capped with a banana leaf, and allowing it to ferment spontaneously at room temperature for two days.
Evaporated milk	Also known as dehydrated milk, evaporated milk is a shelf-stable canned milk product with about 60% of the water removed from fresh milk. It differs from sweetened condensed milk, which contains added sugar.
Eggnog	A drink common during Christmas that contains cow milk, egg and fillers to make the drink quite thick.

Filmjölk	A Nordic dairy product, similar to yogurt, but using a different type of bacteria which gives a different taste and texture from US-type yogurt.
Fromage frais	The name means "fresh cheese" in French (fromage blanc translates as "white cheese").
Fermented milk products	Also known as cultured dairy foods, cultured dairy products, or cultured milk products, fermented milk products are dairy foods that have been fermented with lactic acid bacteria such as Lactobacillus, Lactococcus, and Leuconostoc.
Frozen custard	A cold dessert similar to ice cream, but made with eggs in addition to cream and sugar.
Frozen yogurt	A frozen dessert made with yogurt and sometimes other dairy products. It varies from slightly to much more tart than ice cream, as well as being lower in fat (due to the use of milk instead of cream).
Gelato	The Italian word for ice cream, derived from the Latin word "gelatus." (meaning frozen). Gelato is made with milk, cream, various sugars, and flavoring such as fresh fruit and nut purees.]
Ghee	Ghee is a class of clarified butter.
Gombe	A traditional dish from Sogn og Fjordane in Norway, prepared from curdled unpasteurized milk which is boiled down with sugar for several hours.
Gomme	Gomme is a sort of sweet cheese made of long-boiled milk, having a yellow or brown color. A white, porridge-like variant made of milk and oat grains or rice also exists.
Greek Yogurt	Yogurt that has been strained to remove most of its whey, resulting in a thicker consistency than unstrained yogurt, while preserving yogurt's distinctive sour taste.
Ice cream	A frozen dessert usually made from dairy products, such as milk and cream and often combined with fruits or other ingredients and flavors.
Ice milk	A frozen dessert with less than 10 percent milkfat and the same sweetener content as ice cream.
Junket	A milk-based dessert, made with sweetened milk and rennet, the digestive enzyme which curdles milk.
Junnu	A pudding made by steaming the colostrum of a cow along with Jaggery, Cardamom and optionally Black pepper.
Kaymak	A creamy dairy product, similar to clotted cream. It is made from the milk of water buffalos or cows.
Kefir	A fermented milk drink prepared by inoculating cow, goat, or sheep milk with kefir grains.
Khoa	A milk food widely used in Indian and Pakistani cuisine, made of either dried whole milk or milk thickened by heating in an open iron pan.
Kulfi	A popular frozen dairy dessert from the Indian Subcontinent. It is often described as "traditional Indian Subcontinent ice cream"

Kumis	A fermented dairy product traditionally made from mare's milk.
Lassi	A popular, traditional, yogurt-based drink consisting of a blend of yogurt, water, spices, and sometimes, fruit.
Leben (labneh)	A fermented milk product commonly available in the Arab world, Israel and Cyprus.
Malai	Similar to clotted cream. It is made by heating non-homogenized whole milk to about 80 °C (180 °F) for about one hour and then allowing to cool.
Matzoon	A fermented milk product made from cow's milk (mostly), goat's milk, sheep's milk, or a mix of them and a culture from previous productions.
Míša	A popular Czech confection made with frozen cream cheese.
Mitha Dahi	A fermented sweet dahi or sweet yogurt.
Mozzarella	A heat treated milk that is stretched by hand to expand elasticity.
Mursik	Made from curdled dairy products cooked in a specially made gourd container, it is commonly served at dinner.
Paneer	This fresh cheese is an un-aged, acid-set, non-melting farmer cheese or curd cheese made by curdling heated milk with lemon juice, vinegar, or any other food acids.
Podmlec	Similar to clotted cream, it is made from the milk of goats or cows.
Pomazánkové máslo	This is a spread made from base ingredients of sour cream, milk powder and buttermilk powder.
Qimiq	Consists of 99% light cream and 1% gelatine.
Quark	A fresh dairy product made by warming soured milk until the desired degree of denaturation of milk proteins is met, and then strained.
Qatiq	A fermented milk drink.
Ryazhenka	Fermented baked milk
Semifreddo	A class of semi-frozen desserts, typically ice-cream cakes, semi-frozen custards, and certain fruit tarts. It has the texture of frozen mousse because it is usually produced by uniting two equal parts of ice cream and whipped cream.
Sergem	Made from milk once the butter from the milk is extracted. It is then put in a vessel and heated and when it is about to boil, sour liquid call "chakeu" is added.
Shrikhand	An Indian sweet dish made of strained yogurt.
Skorup	Kaymak that is matured in dried animal skin sacks is called skorup.
Skyr	An Icelandic cultured dairy product, similar to strained yogurt.
Smetana	A range of sour creams from Central and Eastern Europe. It is a dairy product produced by souring heavy cream.
So	Made from layers of milk skin.

Sour cream	Obtained by fermenting a regular cream with certain kinds of lactic acid bacteria. The bacterial culture, which is introduced either deliberately or naturally, sours and thickens the cream.
Spaghettieis	Ice cream made to look like a plate of spaghetti.
Stewler	A fermented milk product made by adding sour cream to baked milk.
Súrmjólk	A cultured milk product, or a type of yogurt. It is made from either whole or semi-skimmed milk and various flavorings are sometimes added.
Tarhana	Dehydrated yogurt and grain product, rehydrated with milk to make soup
Uunijuusto	A dish made from cow's colostrum, the first milk of a calved cow, which has salt added and is then baked in an oven.
Yakult	A probiotic dairy product made by fermenting a mixture of skimmed milk with a special strain of the bacterium Lactobacillus.
Ymer	A Danish soured milk product made by fermenting whole milk with the bacterial culture Lactococcus lactis.
Yogurt	A fermented milk product (soy milk, nut milks such as almond milk, and coconut milk can also be used) produced by bacterial fermentation of milk.
Žincica	A drink made of sheep milk whey similar to kefir. It is a by-product in the process of making bryndza cheese.

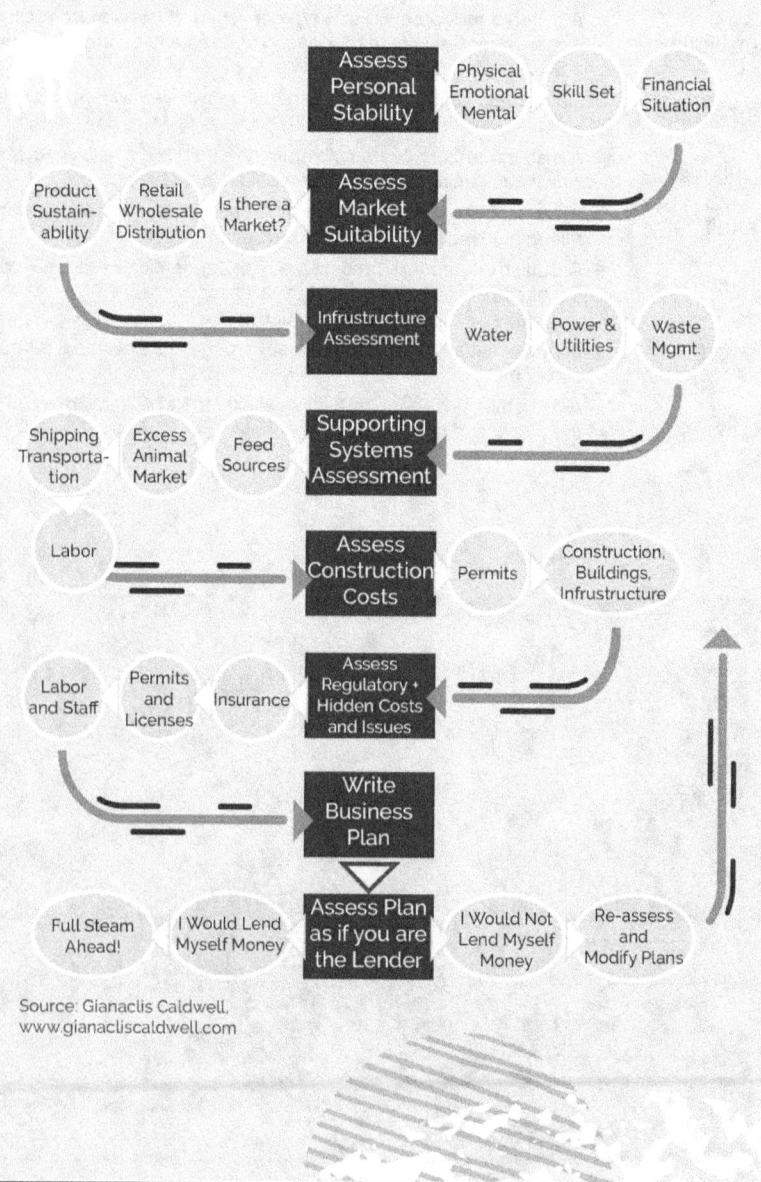

Cheese Making the Really Old-Fashioned Way

Reprinted from Naturally Healthy Goat Guide, Warmke, A. (2012).

Whenever I talk about cheese making I like to say, "I'm a lazy cheese maker." It doesn't make sense to me to make cheeses that require lots of steps in the process. If cheese making is complicated, then I know I won't follow through or enjoy making cheese.

My goal is to use one easy-to-make cheese curd to create a number of different tasting cheese products. With the basic fermented cheese curd it is possible to make a number of different types of cheese products including cottage cheese, spreadable cheese, different tasting semi-hard cheeses plus lemon curd for dessert.

The basic starter cheese is a raw goat milk fermented curd. This type of cheese is a very healthy food with lots of enzymes and vitamins, AND it's easy to make. Plus probiotic!

Terms to know:
- → Curd: The solid protein that separates from the watery part of the milk.
- → Rennet: Comes in either vegetable or animal form. An enzyme from the animals stomach that causes a chemical reaction with the milk to separate the curd from the whey more quickly.
- → Whey: The watery substance that separates from the curd.

What you'll need:
- → Raw milk

- → Soup pot or other large container that holds several gallons of milk
- → Rennet
- → Cotton material for straining the curd (old tee-shirts, reusable jelly bags, tea towels, etc.)
- → Slotted spoon or strainer for removing the curd from the whey
- → A flower or cheese press for allowing the curd to drain (heavy rocks on top of a strainer filled with curd will work also)

Making cheese:
1. Use raw milk - not heated or pasteurized milk. Place the milk in a large container. Dump several gallons into a big soup pot.
2. Add about six drops of rennet (vegetable, animal, lemon juice, vinegar) and stir from the bottom of the pot with a fork - four or five times. At this stage you can add salt, but, if I'm going to add salt I usually wait until the curd has separated from the whey.
3. Let the milk "stand" with a cover over it (to keep out insects and cats) until you can see that the cheese curd has separated from the whey. At any point after the curd has separated you can make cheese. To make a fermented curd, leave the curd and whey together until the curd has bubbles on top of it. Depending on the room temperature, this may take one day or several days.

4. Once the curd has separated from the whey use a sharp knife to cut the curd into chunks.
5. Prepare a strainer by lining it with the cotton material and place the strainer over a container that will hold the whey that will drain out of the curd.
6. Once all of the curd is in the strainer, salt can be added by stirring it into the curd. Herbs or other seasonings can be added at this time as well. Or wait until the curd is completely pressed to the proper dryness, and then place an herb (thyme, basil, mint) on top of the curd and wrap the cheese in wax paper to infuse the cheese curd with the herb flavoring.
7. Either place a weight over the curd to press out more of the whey, or put the curd into a flower press or cheese press to create a drier cheese. Let the curd sit for several hours or over the course of the next 24 hours.
8. Once the curd stops draining, place it in a wrapping of cheese cloth or parchemnt paper or in a glass or plastic covered container - then refrigerate.

What to do with too much cheese:

→ Add different seasonings to make different flavors. Consider adding lemon zest, fresh mint and raw sugar or honey to create a delicious creamy dessert (don't let the curd get too dry for this version).

→ Cut the cheese into small cubes and place in a jar. Fill the jar with extra virgin olive oil or balsamic vinegar and refrigerate. Feel free to add herbs as well. The cheese will be infused with the flavors and last a very long time in this form.

→ Let the cheese dry out in the refrigerator and then grind it up in the blender. Then place it in the freezer so it's won't grow mold, and use the grated cheese on top of pizza, pasta or salad.

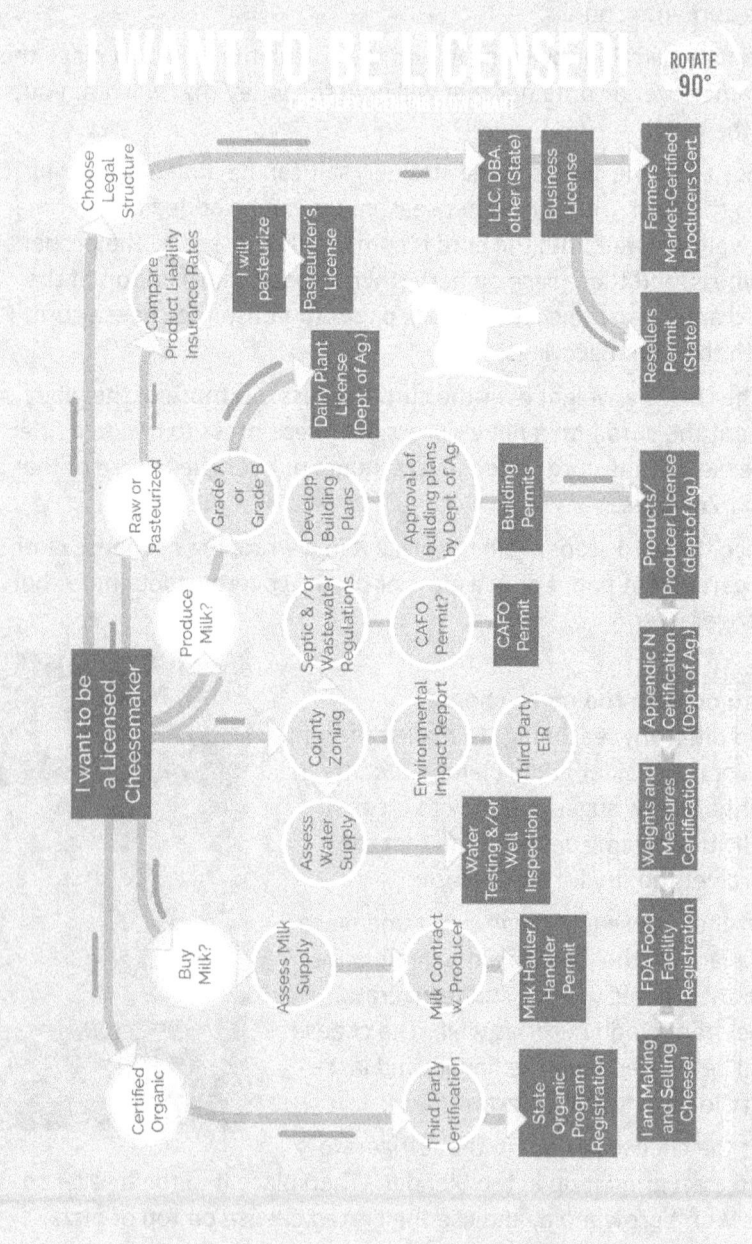

Preparation of Hides for Tanning

An often overlooked by-product of goats are the hides. Whether it's a wether that has been harvested for meat or the unfortunate passing of a milker, why not utilize the hide? A finished hide can be used or sold to make many creative products such as purses, decorative pillows, rugs or clothing.

Online education sources, such as YouTube, can provide the basics on hide tanning. Either tan it at home, or send it to a tannery to have it processed professionally.

How the hide is handled immediately after removal from the carcass is critical. Bacteria is the enemy and can start deteriorating the hide quality, quickly causing hair to fall out.

Buck's County Fur Products in Quakertown, Pennsylvania, specializing in tanning sheep, goats and deer, recommends the following hide preparation prior to shipping.

> *"A finished hide can be used or sold for many creative things, such as purses, decorative pillows, rugs or clothing."*

Preparing Hides for Shipment

Freezing hides in a freezer immediately after harvest is not recommended. This will most often lead to wool loss. Hides that are salted and laid flat are fine in freezing temperatures.

Important: Perform this immediately, body heat can ruin a skin in one hour!

1. **Butchering, removing fat, checking:**
Remove any fat or meat clumps before salting. Thick fat/meat clumps left on hides can cause wool loss. Occasionally a pelt may have a genetic

defect causing a layer of fat to form in the pelt between the two layers of skin.

Cuts around the edges will be trimmed off. Do not send skins with large cuts in the center. Holes will likely cause the hide to tear during tanning.

Talk with your butcher if keeping the hides for tanning. Pulling hides from an animal is best (rather than cutting), as it leaves clean leather without cuts or fat left on the hide.

Make sure to get the hides salted immediately!!! Hides left unsalted can go bad in as little as an hour. The butcher should not throw them in a pile on the floor. Ask the butcher to hang the fresh pelts over a railing, skin side up, until you can pick them up (the same day). Remove any feces clumps, but do not wet the wool.

> *"Make sure to get them salted immediately. Hides left unsalted can go bad in as little as an hour."*

Check hides for extremely matted or felted wool prior to shipping them. The pelt may not be able to be processed if it is felted or matted. Felted fibers commonly result from loosely shed fibers matting into the growing fiber on animals that have never been sheared or in older animals that are butchered in the spring when they are shedding.

2. Salting and drying:
Spread the hides out flat, leather side up, in a protected area. Do not put in the sunshine, as this will burn the leather.

Cover the leather with a thin layer of fine granulated salt. Do not use course or rock salt! Use up to 5 lbs. on large hides, 1-2 lbs. on small hides. Be sure to rub the salt into the edges of the pelt so the edges do not fold over. Use just enough salt to completely cover the hide. Too much salt will draw moisture out of the air, instead of drying the hide.

If moisture puddles up on hide, drain it off and re-salt. Putting the hides on a pallet on a slight incline will help drain moisture. Drying time will vary due to weather factors such as temperature and humidity. The minimum dry time is two weeks. Hides must be completely dry before they can be shipped to the processor. They should not feel wet or damp to the touch prior to shipping.

3. Shipping: Roll the dried hides and put them into a bag or a lined sturdy shipping carton.

NOTE: When salting hides to be tanned, livestock salt from the farm store is the ideal salt for the process.

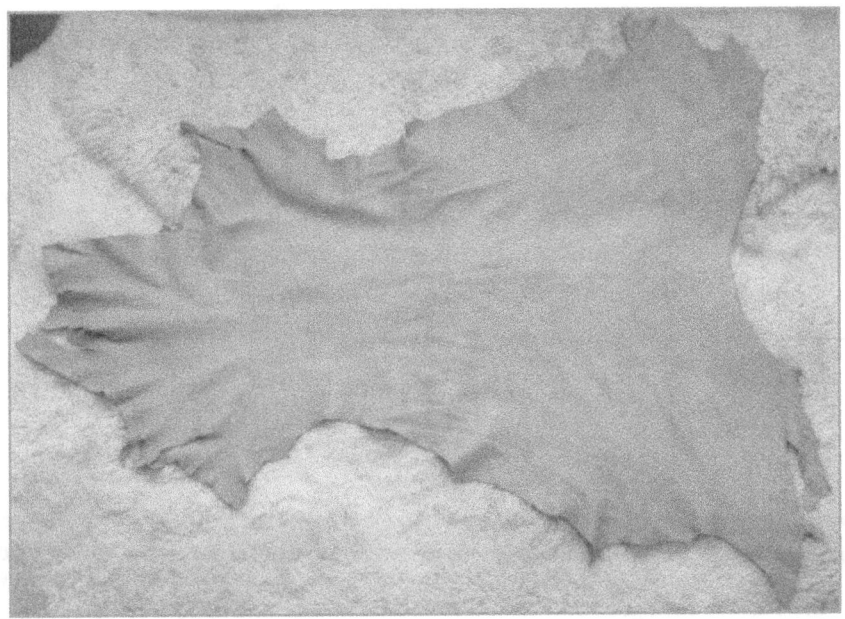

Suggested Resources

→ Boren, J., Baker, T., Hurd, B. & Mason, G. (2004). Tanning deer hides and small fur skins. Guide L-103. New Mexico State University. Cooperative Extension services. Retrieved from https://www.uaex.edu/environment-nature/wildlife/youth-education/TR%20Tanning%20deer%20hide%20AZ.pdf

Woman's Story: Michelle Gorman
Co-owner of Integration Acres and also the Cheese Maker

This whole lifestyle was new to me. I grew up in a big city.

My average day varies depending on the time of the year. In the winter I focus more on the financial part of our business, strategizing for the year, catching up on taxes, focusing on areas I can't get to once kidding season and milk production begins. In March, a typical day starts in the cheese room around 7 am and ends around 3 pm. There are three part time people working for us.

In the last four or five years things have changed significantly because we now have competition with another cheese maker in town. As a result we have started doing wholesale sales to other retailers. It's easier to prepare larger orders and get them out the door.

We work with another local farmer who transports the wholesale products to Columbus, which helps a lot. I used to have to take one day a week to "hit up some businesses," but now I don't have to give up time from the cheese making. Other businesses have been really helpful to us in providing cow's milk when we need it on occasion and helping us purchase inexpensive stainless steel equipment.

We have to stay one step ahead as the market shifts - that's been a challenge. In 2007, locally made cheese was an "oh wow" with consumers - but now it's more mainstream. We analyze the marketplace constantly - we have to if we are going to make this work.

We are always looking at what is going on: Who is the competition? Is the local restaurant scene changing? Is price the issue, or are people willing to pay for quality?

> *"The real challenge is finding the sweet spot where you know it's worth producing the product."*

You might have to think about whether this (cheese production) can be a full time job. If I was only milking goats and making cheese, I don't think it would generate enough money to make a living. We supplement our income by hosting an Airbnb and sell paw paws.

You must always remember that raising goats is a business and sometimes you have to make the difficult decision to cull goats from the herd. An animal that isn't making us any money but we're still feeding it is a liability. It's a tough decision, however, because they're so cute.

We've also invested in young people by training them in our business. We have been most successful with women employees. Our experience with men working on the farm is that they have not been as reliable or as caring with the animals. We pay a good wage and try to share the wealth.

The real challenge is finding that sweet spot where you know it's worth producing the product and then getting people to pay what the product is worth. Goat products are a high end market - yet we live in a very impoverished region. But no matter where you live, there are folks who are willing and able to pay for the product. We just have to find them and create access to the marketplace.

One of the challenges in selling within an artisan market is to educate your customer to the fact that it costs a great deal of money to feed the goats and produce the products they want.

It's fun to work with industry colleagues, but difficult because we all have time constraints. How do you work effectively with your competition? This is a small market, and there are only so many customers.

It's been a great experience for my kids to live on a farm. It is a good life.

> *"Setting prices and calculating costs doesn't have to be rocket science."*

BUDGETING
CALCULATING COSTS: LEARNING HOW TO SET PRICES

Setting prices and calculating costs doesn't have to be rocket science. The formulas and guidelines in this chapter can easily be used to calculate costs and help determine where to set prices that will enable the producer to make a profit.

The first step to good budgeting is good record keeping. Keep all receipts and track spending in a spreadsheet broken down by categories.

Replacement Stock	Grain	Hay	Milking Supplies	Ingredients for Product	Total Cash Required
$ 600.00	$ 1,020.00	$ 832.20	$ 730.00	$ 550.00	$ 3,132.20

	Land Cost	$ 100.00
	Insurance	$ 200.00
Create a simple budget on a spreadsheet	License	$ 230.00
	Market Fee	$ 250.00
	Transportation	$ 157.00
	Total Costs	$ 4,069.20

The spreadsheet can be used to track expenses and can be expanded to easily calculate the not-so-obvious expenses such as mileage to market/feed store/accountant, land cost, taxes, insurance, licenses, market fees, cost for services (like brush hogging or hay making), or the cost of electric to run tank heaters or heaters for barns. These items add up, and they are usually tax deductible*.

> "Keep track of your time. This is important in determining costs."

Infrastructure, capital expenses and depreciation must be accounted for as well. How much did the barn, milking equipment, processing equipment, livestock, tractor (SUV or ATV), fencing, watering systems cost? These expenses are stretched out over their useful life (depreciated), but only a portion of their cost can be deducted each year. State governemtns and the IRS may have different rules for figuring how to deduct these expenses and depreciation.

Those just starting in business won't have the history to accurately budget. A budget is just a "best guess", and even the most experienced producer will find that there are unexpected financial surprises (both good and bad). Use past experience, network with peers, Google prices from potential suppliers and breeders. Don't be afraid to reach out for information!

Keep track of time spent on each task. How much time is spent on daily chores, procuring feed, sourcing new animals, breeding and kidding, driving to pickup, or to deliver product to customers or to the market?

Also, how much time is spent milking, creating your product, marketing (social media, emails etc.) and time spent interacting with customers? This information is important in determining costs. Time, even the time of the owner/farmer, has value and should be accounted for when setting prices.

PRICING STRATEGIES:

So now that cost of production has been determined, it is time to calculate the cost of the product. It is important to make profit or return on the investment or the farm/business will not be sustainable long-term. Price setting seems to be one of the greatest struggles for many farmers.

"Do not apologize for pricing your products in such a way that all costs are covered and you have a return that will make you want to stay in business for the rest of your life." Joel Salatin

> *"Many farmers simply look at what their competitors are charging and base their prices from that."*

Many farmers simply look at what their competitors are charging and base their prices accordingly. This is called "market pricing," or sometimes referred to as competitive pricing. In this model, examine the prices of as many other similar products as possible and average them to set a retail price.

Another option used when setting prices is to determine and allocate all costs, then add a profit margin. This pricing system, called "cost plus mark up" will ensure a profit, but may result in prices that are not competitive with other providers of similar products.

Perceived value pricing is another pricing option. This pricing method bases the product price on the effective value to the customer, relative to alternative products in the marketplace. If the farmer has done a good job in convincing the the customer of the unique value of a product, a higher profit margin may be possible.

What is the value added by having certain products available to customers such as Tuberose Soap? Although the essential oil to make it is more expensive, and the profit less per bar, it has the potential to bring customers to the farm store or booth at the farmers market and they buy other things as a result.

> *"What is important is that the venture is worth it to you in a combination of money and experiences."*

All three pricing models may be used whether the product is sold retail or wholesale. Obviously the price charged for wholesale sales must be lower to ensure the retailer has the ability to make a profit as well. The reduced costs involved in not having to provide a retail outlet, as well as the high volume are factors in allowing for the lower price while still making an adequate profit.

Remember that not all income is monetary. Goat farmers produce food and other products that are consumed by their family. How much would that have cost to buy at the market?

How much would all the physical exercise involved in raising goats have cost? (think gym memberships, improved health). What did that time spent caring for those goats do for personal happiness? Could that be a spiritual practice? If children have been enlisted to help or observe... what educational experiences did they receive? What is the value of that?

Barter is a great way to lower costs. Hay, accounting services, a brush hog, vegetables and other food can often be traded for goat products. Just remember, from a tax standpoint, the value of traded items are still counted as income. For recordkeeping purposes use whatever dollar amount you would have received when selling the bartered goods.

It is important that the venture is worth the combination of money and experiences.

*NOTE: This book is not offering tax advice; just describing past practices. Be sure to speak with an accountant or check the tax codes.

SUGGESTED RESOURCES

→ Salatin, J. (1998). *You can farm: The entrepreneur's guide to start and succeed in a farm enterprise.* Polyface.

→ Ayers, A. (2018, March 3). How to price your products - with a FREE pricing calculator. Retrieved from https://www.launchgrowjoy.com/how-to-price-your-products/

→ Entrepreneur Small Business Encyclopedia. (n.d.). Pricing a Product Definition. Retrieved from https://www.entrepreneur.com/encyclopedia/pricing-a-product

→ Posek, S. Maker's Row Blog. (2014, June 17). How to Calculate Retail Prices. Retrieved from https://makersrow.com/blog/2014/06/how-to-calculate-retail-prices

→ Small Food Business (2013, February 15). Understanding the difference between markup and margin. Retrieved from http://www.smallfoodbiz.com/2013/02/15/understanding-the-difference-between-markup-and-margin

Hay Budgeting Worksheet

Non Milking Goats

# of goats	approx. total wt. (lbs)	total weight (lbs)	4% body weight (X 0.04)	=	Hay in pounds needed per day (lbs)

Milking Goats

# of goats	approx. total wt. (lbs)	total weight (lbs)	6% body weight (X 0.06)	=	Hay in pounds needed per day (lbs)

$ price of bale	/ weight of bale (lbs)	= $ cost per pound		$ cost per pound	X pounds per day (lbs)	= $ cost per day

EXAMPLE

Non Milking Goats

# of goats	approx. total wt.	total weight	4% body weight	Hay in pounds needed per day
3	150 lbs	150 lbs	X 0.04	6 lbs

Milking Goats

# of goats	approx. total wt.	total weight	6% body weight	Hay in pounds needed per day
4	280 lbs	280 lbs	X 0.06	16.8 lbs

price of bale	weight of bale (lbs)	cost per pound		cost per pound	pounds per day	cost per day
$5.00	/ 50 lbs	= $0.10		$0.10	X 22.8lbs	= $2.28

Woman's Story: Abbe Turner
Abbe Turner, Owner of Lucky Penny Farm in Garretsville, OH

Since 2006, I have been living with goats at Lucky Penny Farm, where I am a cheesemaker and goat herder. I use goats' milk to make artisan cheeses, soaps, and cajetta - a traditional, caramel goat-milk sauce.

The small farm dream was a part of my world much earlier than it became a reality. My background was in professional fundraising and economic development. In my former office I had a photo of a beautiful pastoral farm and that was my focal point.

I kept the photo in a place where I could see it multiple times a day and would look at it and know, "This is why I am working. To eventually have that farm."

In 2002 I found a lovely historic property in Garretsville, OH. It was an 1870 farmstead with five out-buildings and a lot of work to do.

At the time I had just started a young family, having a two and three-year-old. Not having experienced livestock agriculture in my upbringing, I knew I did not want to start with cows, but I loved cheese, so I turned to the Farm and Dairy section of the newspaper, and discovered that goats were available. I picked up the phone, bought four goats, and only later told my husband. That was the beginning.

The goats provide a wonderful rich milk, which is of course delicious for drinking, but also great for making really lovely cheeses that highlight Ohio's wonderful soils.

The dairying business is an interesting business. It is 24/7, 365 days of the year. There is a seasonal flow, which means there will be times when there is more milk than there is demand and vice versa.

I have discovered that making cajetta (the goat-milk caramel sauce) is a great way to take any excess and turn the goats' milk from having a two-to-three week shelf life as a cheese, to something that is shelf stable for about a year.

> *"Being successful is the result of a lot of trial and error - constantly evaluating what is working and what is not."*

Cajetta was originally included in my business plan as a way to balance the seasonal flow of increase and decrease in goats' milk, as well as balance orders and demands from the market place.

I have had many different business models over my many years of farming and originally planned to simply make a farmstead cheese. However, the economics of that model did not entirely work out.

Back in 2009, when my partner and I were planning to go to market, the economy was in a recession and as a result we could not obtain traditional financing.

As a result, we tried lots of value-added approaches, such as: aging cheeses, marketing our sustainable farming practices, and making sheep cheese (which has a shorter lactation cycle than goats).

Being successful is the result of a lot of trial and error - constantly evaluating what is working in the marketplace and what is not.

Dairy as a business has extremely high entry costs. Because of regulations, one must build a commercial creamery which costs upwards of six figures.

Women farmers are typically care takers. They provide a greater attention to the methods of production. They tend to raise animals in a non-traditional (if factory farming can now be considered "traditional") manner, focusing on grass-fed or organic models for example.

> "Women are demonstrating different methods of production."

Women are demonstrating different methods of production and allowing people to farm in the way that they want to farm—whether that's using only heirloom seeds or biodynamically. We need to support these choices in our purchases. We need to place a value on these methods of production - supporting what women farmers do with their hands as well as with their hearts.

When I began writing my first business plan in 2006, I went to talk to a woman who was getting out of the business. The woman suggested I run like hell!

However, I believe that if you have a calling to do something, you should work every day to get closer to making that plan a reality. Every day do something to get closer to your goal. "Just do it! Grab a friend, reach out and talk to everybody you can. Do it."

I have found many great partners in the Amish community and with chefs from the greater Cleveland area. I see myself as building a bridge between the rural and urban areas; teaching consumers to reflect upon how our farmers are farming and how our food effects our bodies.

THE ART OF SIMPLE BRANDING AND MARKETING

Branding is defined as the marketing practice of creating a name, symbol or design that identifies and differentiates a product from other products. Creating a marketing plan can seem daunting, but it needn't be. Here are some simple and affordable ways to help create a brand.

> *"Make a connection with your customers. Consider how they relate to your brand."*

Get personal. Make the name and logo tell a story to draw the customer in. A personal connection will help people remember the brand. Grow the brand by linking to memorable phrases, images and taglines.

Make the first impression count, don't waste time or money on homemade business cards and poorly designed literature. Professional looking business cards can be purchased online for about the cost of the ink and paper required to make them at home. Some suppliers even offer matching t-shirts, banners, and bags for reasonable prices to create that whole brand image.

Humor can be a good icebreaker and can help customers remember the brand, but the business owner should be careful not to be offensive. Ask friends and family to evaluate logos and materials to check for potential bloopers.

Make a connection with potential customers, consider how they relate to the brand. Is it immediately apparent what the business is trying to convey? Use style and content that is appropriate for the line of business. If the business is selling goat milk soap, a "country" feel to the logo may be appropriate. For example, a gingham background with a symbol that clearly indicates "goat."

Consider the cultural and sociological significance of the brand. Does the brand have subliminal recognition? From the logo can someone immediately tell what product is being sold?

If there is not a large budget for logo design, utilize students for logo design. Many students are trying to build a portfolio and even just a bit of extra cash is always beneficial.

Also be concious of how the logo will look in various sizes and colors. The logo should look as good on a business card as it does on a banner over a booth at the farmer's market. Also, logos that are dependant upon color to be effective may fall flat if the marketing piece is reproduced in black and white.

Don't baffle customers or overload them with too much information. Use the KISS method, "*Keep It Simple Silly*". Test and measure the brand and message. Everyone will have different opinions, but try to get a general consensus. The one who decides if the message is clear is the customer, not the business owner.

Focus on who is being targetted and why. Clear design doesn't drown out the message with too many words or unrelated imaging. Make the brand benefit clear and obvious. Don't be afraid of white space, less is more in some cases.

Maintaining consistent messaging promotes the feeling of dependability and security with customers - making it easier to trust the brand. Does the business' website look like it belongs with its business cards? Does marketing literature convey the same feel and message as their banners and signage? Is the booth in the same spot at the market every week? Consistency also helps find and remember the brand.

The business should aim to set itself apart from its competitors. Check out the competition and identify what they are doing right as well as what they are doing wrong.

A brand should not be afraid to let customers know why it is better than the competition - but care should be taken not to be seen as being critical or "badmouthing" the competition. Focusing on the positive benefits of the brand is always a better strategy than pointing out the deficiencies of the competition. Let the customer make that connection.

> *"At the market a business typically has only a few seconds to engage a potential customer, make it count."*

Get the message out! Once a brand is established, spread its message around.

Create a ten second or less "elevator pitch" that explains the business instantly, then try it out on five or more friends. A business typically has only a few seconds to engage a potential customer, make it count and draw them in.

Offer to speak at local clubs such as Rotary, Mother's Clubs, Kiwanis, FFA etc. They are always looking for fresh content at their meetings and it's a great way to get buy in and build relationships.

If the farm offers agritourism or on-site sales, create a map of interesting things to do in the area that will make potential customers feel the trip is worthwhile. While they are visiting the region they can also check out that "Giant Ball of String Park" or the local zoo.

Invite local businesses to visit and discuss cooperative marketing and promotion opportunities that will be mutually beneficial. This also helps in networking and gets buy-in from neighbors.

Take advantage of the benefits of Point of Sale. Use the POS to track all sales, not just credit card sales. There is a lot of information that can be gleaned from the data, such as: the days of the week that have the best sales, what times are the best at the farmer's market, who the best customers are. Create professional looking invoices, have an online store, create a rewards/loyalty program, even send coupons. Experiment with tools that help maintain a relationship with customers, identify successes and help correct those areas that need tweaking.

Of course in today's market, social media as an important tool. A photo is worth a million words and the cuter the picture the more interaction it will receive. Frequent sharing will help keep customers engaged and feeling like they are part of the team.

"Keep customers engaged and feeling like they are part of the team."

Keep the audience in mind and make sure any content (especially photos) are appropriate to the brand. Look for free social media marketing training at a local library or ask the farmer's market association to offer social media training.

Crowdfunding

Need funding for a farm business or a project? Perhaps sharing the process of the project would be helpful to others. The Internet provides a number of opportunities to share the risk of a project or obtain funding through non-traditional sources. Rather than go into debt, many farm businesses are turning to crowdfunding to obtain capital for specific projects or to expand a business.

The idea of raising money online may seem that it would require the eating of humble pie, but studies show that the number one reason people don't give to causes is.... drum role please... that they weren't asked.

> *"Crowdfunding online is one of the fastest growing ways businesses are raising startup capital today."*

Success is often as simple as giving people the opportunity to invest in an idea. Amounts contributed typically range from five dollars to thousands of dollars. However, some campaigns sweeten the pot when asking for funding by creating different levels of giving - with rewards for the contributions at the various levels. For example, if someone gives $50, they might receive a bar of homemade goat soap. If they give a $1,000, they might receive a stay on the farm with two friends.

Growth of Crowdfunding

The Crowdfunding Industry Report by Massolution put out data showing the overall crowdfunding industry has raised $2.7 billion in 2012, across more than 1 million individual campaigns globally. In 2013 the industry grew to $5.1 billion.

In late 2017 the World Bank predicted that crowdfunding would continue to grow at an amazing rate, topping $93 billion worldwide by 2025.

Crowdfunding Models

There are two main models or types of crowdfunding. The first is what's called **donation-based funding**.

The birth of crowdfunding evolved from this model, where funders donate via a collaborative goal-based process in return for products, perks or rewards.

The second and more recent model is **investment crowdfunding**, where businesses seeking capital sell ownership stakes online in the form of equity or debt.

> *"The various crowdfunding sites feature different models and focuses."*

In this model, individuals who fund become owners or shareholders and have a potential for financial return. This is unlike in the donation model, where funders are not owners and will not participate in the success or failure of the program.

Crowdfunding Sites From Which To Choose

Business owners normally use different crowdfunding sites than musicians. Musicians use different sites from causes and charities. Below is a list of crowdfunding sites that have different models and focuses. This list can help in finding the right place to meet crowdfunding goals and needs.

→ **Kickstarter**
Kickstarter is a site where creative projects raise donation-based funding. These projects can range from new creative products, like an art installation, to a cool watch, to pre-selling a music album. It is not for businesses, causes, charities, or personal financing needs. Kickstarter was one of the earliest platforms, and has experienced

strong growth and many break-out large campaigns in the last few years.

→ **Indiegogo**
While Kickstarter maintains a tighter focus and curates the creative projects approved on its site, Indiegogo approves donation-based fundraising campaigns for almost anything -- music, hobbyists, personal finance needs, charities and whatever else one could think of (except investment). They have had dramatic international growth because of their flexibility, broad approach and their early start in the industry.

→ **Crowdfunder**
Crowdfunder.com is the premier platform for raising investment (not rewards) capital, and has a one of the largest and fastest growing network of investors. Success stories such a a feature about a company raising $2 billion on the site have gained the site a lot of media attention. After getting rewards-based funding on Kickstarter or Indiegogo, companies often give investors an opportunity to invest at Crowdfunder to raise more capital.

→ **RocketHub**
RocketHub powers donation-based funding for a wide variety of creative projects. What's unique about RocketHub is their FuelPad and LaunchPad programs that help campaign owners and potential promotion and marketing partners connect and collaborate for the success of a campaign.

→ **Crowdrise**
Crowdrise is a place for donation-based funding for causes and charities. They have attracted a community of "do-gooders" and fund all kinds of inspiring causes and needs. A unique points system on Crowdrise helps track and reveal how much charitable impact members and organizations are making.

→ **Somolend**
Somolend is a site for lending for small businesses in the US, providing debt-based investment funding to qualified businesses with ex-

isting operations and revenue. Somolend has partnered with banks to provide loans, as well as helping small business owners bring their friends and family into the effort. With their Midwest roots, a strong founder who was a leading participant in the JOBS Act legislation, and their focus and lead in the local small business market, Somolend has begun expanding into multiple cities and markets across the US.

→ **appbackr**
For businesses hoping to build the next new mobile app and are seeking donation-based funding to get things off the ground or growing, then check out appbackr and their niche community for mobile app development.

→ **AngelList**
Tech startups with a shiny lead investor already signed on, or looking for Silicon Valley momentum, will find a number of "angels" and institutions seeking investments through AngelList. For a long while AngelList did not market themselves as a crowdfunding site, which makes sense, as they catered to the investment establishment of VCs in tech startups. But now they're getting into the game. The accredited investors and institutions on AngelList have been funding a growing number of top tech startup deals.

→ **Invested.in**
For those wishing to create their own crowdfunding community to support donation-based fundraising for a specific group or niche in the market, there is Invested.in. Invested.in is a Venice, CA based company that is a top name software provider, providing the tools to get started and grow an individualized fundraising program.

"Crowdfunding is growing a market for impact investing in social enterprises."

→ **Quirky**
Inventors, fabricators, or tinkerers of some kind or other may find that Quirky is a place to collaborate and crowdfund for donation-based funding with a community of other like-minded folks. Their site digs deeper into helping the process of bringing an invention or product to life, allowing community participation in the process.

"Crowdfunding is accelerating angel investing and creating an entirely new market for investment."

How Crowdfunding Is Shaping A New Economy

Crowdfunding has revitalized the arts at a time when the public programs that have traditionally supported them are steadily dying off.

Crowdfunding is growing a market for impact investing in social enterprises, marrying the worlds of entrepreneurship and philanthropy, and helping a broader base of investors to back companies for both profit and purpose.

Crowdfunding is accelerating angel investing and creating an entirely new market for investment crowdfunding for businesses.

So get involved and join a crowdfunding community today. The use of this fundraising tool can make a huge difference for a project or business, and also help build a new and more collaborative economy.

SOCIAL MEDIA

The term "social media" can be confusing. The world of Facebook, Instagram, Twitter and other social media sites can seem complicated.

Despite an intial reluctance, the world of agriculture needs these platforms to easily promote information on products, and to market services. The cost of marketing through social media is usually free or quite inexpensive. Learning how to use the simplest version of the social media sites takes only a little time and effort, but the payback can be very powerful when "friends" share the information on the product with their friends and contacts. Social media is the modern version of US postal mass mailings... and it works.

> *"The cost of marketing through social media is usually free or quite inexpensive."*

Social media is defined as an electronic communications channel dedicated to community-based input (friends, family, acquaintances), interaction (personal messages), content-sharing (news, photos, and event information) and collaboration (blogs and forums). Websites and applications dedicated to forums, microblogging, social neworking, social bookmarking, social curation and wikis are among the different types of social media.

INSTAGRAM
(Reprinted from The Cheat Sheet, Jan. 22, 2018)

Instagram is becoming an increasingly popular and powerful social network. The Facebook-owned photo sharing app is still significantly smaller than Facebook, still the most ubiquitous social network in the world. But Instagram is growing much more quickly than Facebook, as new users join everyday to connect not only with their friends, but with a global community of users who share photos and videos from their everyday lives on the platform

How to Sign Up for Instagram
To sign up for Instagram, you need to download the mobile app on an iOS or Android phone or tablet, or on Windows Phone 8 or above. (Download Instagram from the iOS App Store, from the Google Play Store, or from the Windows Phone Store.) While you can view Instagram online, you'll need to create your account using the Instagram app.

> *"Instagram is growing much more quickly than Facebook."*

Once you've downloaded the app, tap the Instagram icon to open it. Tap "Register with Email" to sign up with your email address, or choose "Register with Facebook" to sign up using your Facebook account. If you register with an email address, the app will prompt you to create a username and password and fill out your profile information. (Once you've finished that, tap "Done.") If you choose to register with Facebook, the app will prompt you to sign in with your Facebook account if you're currently logged out.

If you want to edit your profile information, such as your name, username, or email address, you can go to your profile by tapping the person icon in the toolbar at the bottom of the screen. Tap "Edit Your Profile" and type your new name, username, website, or bio and tap "Done" on the iPhone, "Save" on Android, or the checkmark on Windows Phone.

How to Navigate Instagram
Your profile shows your bio and the photos and videos you've posted to Instagram. From your profile, you can also edit your profile information and adjust your Account Settings. You can navigate to your profile by tapping the person icon in the toolbar at the bottom of the screen, and make changes by tapping the "Edit Your Profile" button or access more information and choices by tapping the gear icon to access Options. You

can write a bio of up to 150 characters on your profile, or add or change a profile photo to import from your phone's library, Facebook, or Twitter.

The camera enables you to take photos with the Instagram camera or share photos from your phone's photo library. You can access the camera by tapping the center icon in the toolbar at the bottom of the screen, and from there can take a photo or video or choose a photo from your photo library.

You can use the Search & Explore function to find people to follow, search for specific users, and explore hashtags. Access Search & Explore by tapping the magnifying glass icon in the toolbar at the bottom of the screen, and toggling between the "Photos" and "People" views to explore posts and users, or type in the "Search" box and choose between "Users" and "Hashtags."

> *"Hashtags will help other Instagram users to find your posts."*

How to Take or Post a Photo with Instagram

When you share posts on Instagram, you have two options: taking a photo with Instagram's camera, or uploading one from your phone's photo library. You can tap the camera icon at the bottom of the screen to take a photo, or to select one from your phone. From the camera, you can switch the photo grid on and off, switch between the front-facing and back-facing camera, or choose whether or not to use the flash. You can tap the shutter button to take a photo, or hit the video icon to record video.

Once you've chosen a photo, you can add effects from the editing tools at the top of the screen or select a filter from the choices at the bottom of the screen. Then you can write a caption for the photo, tag people in the photo, add it to your Photo Map, or share it to Facebook, Twitter, Tumblr, or Flickr. When you write the caption to your photo, you can use #hashtags and @mentions. Mentions let you bring a post to the attention of another user — who will get a notification when you mention him or her in a post — and hashtags will help other Instagram users to find your posts, or help you to tag or categorize your posts for yourself.

To tag people in a photo, tap "Tag People" and tap on someone in the photo. Start entering their name or username, and select the correct users from the drop down menu that appears. If the person that you want to tag doesn't appear in that menu, tap "Search for a person" to find him or her. If your photos are public, then anyone can see the photo, and the person tagged in the photo will get a notification. If your photos are private, then only people who are following you will be able to see the photo. The person whom you tag in the photo will get a notification if they're following you. (You can see and manage the photos that other people tag you in from your profile.)

When you post your photo, it will appear both on your profile and in your feed. If you've set your profile to private, only the people whom you've approved to follow you will be able to see it.

How to Record or Post a Video with Instagram
To take a video with Instagram, tap the camera icon and tap the video icon to switch from photo mode to video mode. Press and hold the video icon to start recording, and lift your finger off the button to stop recording. Tap "Next" to add a filter and then share your video, exactly as you would with a photo that you've taken with Instagram or selected from your phone's photo library. (But it's worth noting that video recording is currently only available in the iOS and Android versions of Instagram, and isn't yet an option for Windows Phone users.)

To share a video that you've previously recorded with your phone, tap the camera icon and then tap the video icon to switch from photo mode to video mode. Tap the box in the bottom right to view your phone's video album. Choose the video that you want to upload, and if you're on an iPhone, press "Next." Then, choose which 15 seconds of the video you want to share by placing your finger on the strip at the bottom, and sliding to select where the clip begins playing. Then, drag the blue slider above the video strip to choose where the clip ends. Tap "Add" on the iPhone or the arrow on Android.

TWITTER
(Reprinted from Wired, Aug. 29, 2018)

Twitter is where news is broken (the first to report an event), links are shared, and memes are born. It's also a place for chatting with friends. Yet unlike Facebook, Twitter is public by default. And that's not a bad thing. It means your jokes can go viral (if they're funny) and in addition to your friends, you can interact with your favorite journalists, athletes, artists, or political figures, all in the same space.

Generally speaking, tweets show up in the order they happen. At the top of a Twitter feed, you'll see tweets that are only a second old. New tweets appear at the top, pushing the older ones down. If you haven't signed on in awhile you might get a box of recommended tweets you may have missed, but outside of that the equation is simple: The further down you scroll, the older the tweets get.

> *"Twitter is where news is broken, links are shared, and memes are born."*

Tweets can contain links, photos, GIFs, or videos. But if you're tweeting text, you're limited to 280 characters. It used to be 140, which was even more stifling, but once you get used to it you'll learn to love the brevity. It helps make your tweets pithy, and there's much less rambling you have to read when scanning other tweets.

How to Make a Twitter Account on Desktop

Step 1: Go to Twitter.com or download the app and sign up for an account. The "Full name" that you provide will be your display name, but unlike Facebook, you can change your display name to whatever you want as many times as you want, so it's really easy to stay anonymous if you so choose.

Step 2: Enter in your phone number. This is a form of authentication that will help in case you ever lose access to your account. You'll want to use a phone you actually have access to because the next step will ask you to verify a number sent via text.

Step 3: Pick a password, and make it secure! You don't need a troll getting a hold of your account and dismantling the reputation you've worked so hard to build.

Step 4: Choose your interests. This will help with the next step, which is where Twitter will give you suggestions of people you can follow. You can also skip both of these by saying "skip for now" in the top right hand corner.

Step 5: Once you're in, click on the grey silhouette next to the "Tweet" button on the top right of your screen, and click "Settings & Privacy." At the top, you can pick a username you like. That will be your username, or handle, and people can notify you by typing @ in front of your username in a tweet. Choose something you like that you think isn't taken, but also something easy to remember for others.

Step 6: Pick an avatar. The default picture is a silhouette, but you can make your avatar whatever you want (your face, a dog on a skateboard, the possibilities are endless). Just click the silhouette and head to "Profile" and then click "Edit Profile" on the right underneath the blue bar. You can update your header photo from this place, too. Be sure to read Twitter's rules for avatar images to make sure what you pick is not in violation.

Step 7: Write your bio. You may wish to list where you work, live, or a line from a favorite poem in your bio. This is the short blurb that lets potential followers know who you are and what you're likely to tweet. There is also a handy spot to list your website, if you have one.

FACEBOOK
(Reprinted from Lifewire, June 25, 2018)

Facebook is the Internet's most widely used social network, with nearly 1 billion people using it to connect with old friends and meet new ones. People use Facebook to create personal profiles, add other users as "Facebook friends" and share information with them in various ways. How Facebook works can be a bit mysterious to new users, but it's all about communication, so learning the network's core communication tools is essential.

After signing up and adding friends, people communicate with some or all of their Facebook friends by sending private, semi-private or public messages. Messages can take the form of a "status update" (also called a "post"), a private Facebook message, a comment about a friend's post or status, or a quick click of the "like" button to show support for a friend's update or a company's Facebook page.

"Facebook is the Internet's most widely used social network, with nearly 1 billion people using it."

Once they learn Facebook, most users share all kinds of content - photos, videos, music, jokes, and more. They also join Facebook interest groups to communicate with like-minded people whom they might not otherwise know. After growing familiar with how Facebook works, most people also use special Facebook applications that are available to plan events, play games and engage in other activities.

Signing Up:
The first step in using Facebook is to sign up and get a new Facebook account. Go to www.facebook.com and fill out the "Sign Up" form on the right. You should give your real first and last name along with your email address and the rest of the form. Click the green "sign up" button at the bottom when you're done.

Facebook will send a message to the email address you provided with a link asking you to confirm your email address. You'll need to do this if you want full access to Facebook's features.

> *"Facebook calls its profile area your Timeline, because it arranges your life in chronological order."*

If you're signing up to create a business or product-related page on Facebook, click the link below the sign-up form that says "create a page for a celebrity, band, or business" and fill out that sign-up form instead.

After signing up for Facebook, skip the next part where it asks to import your email contacts to help build your friend list. You can do that later. First, you should fill out your Facebook profile before you start connecting with many friends, so they'll have something to see when you send them a "friend request."

Facebook calls its profile area your "Timeline" because it arranges your life in chronological order and displays a running list of your activities on Facebook. At the top of the Timeline is a large horizontal banner image which Facebook calls your "cover" photo. The inset below it is an area reserved for a smaller, square "profile" picture of you. You can upload the image of your choice; until you do, a shadowy avatar will appear.

Your Timeline page is also where you can upload basic biographical information about yourself--education, work, hobbies, interests. Relationship status is a big deal on Facebook, too, though you don't have to publicize your relationship status if you don't feel like it. This Timeline/profile area is where other people will go to check you out on Facebook, it's also where you can go to check out your friends because each of them has a Timeline/profile page.

After filling out your profile, you can start adding friends by sending them a "friend request" via an internal Facebook message or to their email address if you know it. If they click to accept your friend request, their name and a link to their profile/Timeline page will automatically appear on your list of Facebook friends.

Facebook offers various ways to find friends, including a scan of your existing email contact list if you grant access to your email account. Searching for individuals by name is another option.

As soon as you have a few friends and have "liked' some companies, comments or products, then Facebook's automated friend recommendation tool will kick in and start showing you links to "people you may know." If you recognize their face when their profile image appears on your Facebook page, you can just click the link to send them a friend request.

SUGGESTED RESOURCES

- → Bolluyt, J. (2018, January 22). How to Use Instagram. Retrieved from https://www.cheatsheet.com/gear-style/how-to-use-instagram.html
- → Staff, W. (2018, August 29). How to Use Twitter: Critical Tips for New Users. Retrieved from https://www.wired.com/story/how-to-setup-twitter-search-hashtag-and-login-help
- → Walker, L. (2018, June 25). Learn Facebook tutorial - How Facebook works. Retrieved from https://www.lifewire.com/tutorial-how-facebook-works-2654610

GOAT GUIDE FOR LIFE

1. Make it their idea
2. Everybody needs a mama
3. Lead without fear
4. Don't be afraid to challenge the herd
5. Know when to come in out of the bad weather
6. A little dose of rest in the sunshine cures lots of ills
7. Rest when you need to, no matter what is going on
8. Know who is the leader, and follow her
9. Try a little of this and a little of that
10. Don't stick your nose where it doesn't belong
11. Teats rule
12. Massage is the answer to good health
13. Keep your horns
14. Expect the unexpected
15. Don't make assumptions
16. Take care of your feet
17. A smile brings a smile
18. Use your horns for good
19. Be curious
20. Try everything once
21. Eat real food
22. Don't eat corn
23. Don't be afraid to challenge the herd
24. Be loud...be proud
25. Poop is valuable - it all makes soil
26. Good touch produces results

© Blue Rock Station LLC

Woman's Story: Annie Warmke
Co-owner of Blue Rock Station, Farmer, Goat Herder, writer

Blue Rock Station (BRS), a sustainability homestead near Zanesville, Ohio, is a place I have been walking to my whole life. One of the best parts of life at BRS is being a goat herder. I love it so much that I started a Goat College program in 2010, because I wanted people to share some of the basics in natural care of goats. Goats are amazing.

My goat herding life began when I attended a workshop where It occurred to me that raw milk could be something I could provide for my family to promote health, plus it seemed to incorporate my love of animals, learning, and natural products. I started slowly by buying raw goat milk from a local farmer.

Within two months I bought Eleonore Rigby and her two kids (Sanaan and Alpine mix). After lots of mistakes and trial and error, the goat herd at BRS has grown from the original three to as many as 20. The bucklings are banded a few weeks after birth to make them wethers. The wethers are worked with so that they can be adopted for a variety of placements, such as: a petting zoo, personal pets, for brush eating, working with specific groups of children, and packing (hiking goats). The doelings are either sold after six weeks or bred at 80 lbs. to produce a milk goat for an individual family, or for the basis of a new herd.

As a young girl I remember my grandmother, who was a country woman had a needlepoint pillow that read, "Bloom where you are planted." With hard work and a plan, my herd and I have done just that in the hills of Southeast Ohio.

Animal Health Issues

Goats are ruminant animals. Their digestive tracts, which are similar to those of cattle, sheep and deer; consist of the mouth, the oesophagus, four stomach compartments, small intestine and large intestine.

Like other ruminant animals, goats have no upper incisor or canine teeth. Goats depend on the dental pad in front of their hard palate, lower incisor teeth, lips and tongue to take food into their mouths.

How Digestion Works

Goats have a four-chambered stomach consisting of the rumen, the reticulum, the omasum, and the abomasum. Each chamber performs a different task.

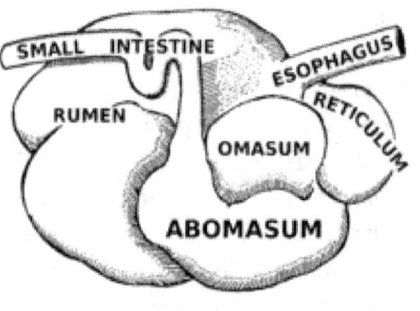

These chambers work like fermentation vats, using bacteria to break down the food. A goat's diet must have an ample supply of fiber and roughage in order for the rumen system to work properly. Changing a goat's diet too suddenly will disrupt the balance of bacteria and can be very dangerous. A healthy diet for a goat will be very stable with few dramatic variations from day to day.

Basic Philosophy of Health

Goats have only three basic health challenges (besides age, of course) that impact their ability to maintain health. The first is bacteria and parasites. The second is disease. The third is injury.

A mammal's immune system is the basis of health. If the food, water and environment provided for the animal speaks to and supports the immune system, then the efforts to maintain health, and productivity will be successful most of the time.

Hippocrates is quoted as saying, "First, do no harm." Good advice. All the treatments suggested in this book will not harm a goat - even if the diagnosis is wildly off base. It is best to take the "low-tech, high-touch" approach here, plus look at pedigree vs basic health maintenance.

Also, it is a sad reality that, at least in the United States, there may be little alternative but to treat the animal yourself. As the U.S. veterinary system moves ever closer to the medical model - it is often a case of simple economics and too many drugs.

No wonder many farmers simply "put down" or butcher injured or sick goats, rather than treating them.

Sometimes (obviously) there are accidents, particularly with the goat's need to inspect, but if the goat has a strong immune system, the goat will have a better chance of overcoming any injury and be better able to regain health.

> *"The sad reality is that there may be little alternative but to treat the animal yourself."*

Each animal has a handy built-in system for healing, creating immunity, keeping healthy, producing (milk or offspring), and growth. The goal is to tap in to this wealth within the animal so that they can do their job.

At the same time, the goat herder must provide the goat with the tools they need to support this natural system by making sure that they receive the right mix of food (and vitamins/minerals), clean water and a healthy environment. This partnership can be a winning combination – the animal has what it needs to live a long healthy productive life, and the humans receive food, income and healthy offspring.

Most goat herders look only at an animal's conformation and pedigree. They want to know how much milk the ancestors gave, or if the udder is lopsided.

These factors do matter, but what is this animals "peaceagree"? Can they get along in the herd? What will their contribution be to the herd and to the farm? Will they be good consistent milkers and produce healthy kids? Will they be able to live a long productive life, and not need to be sent away or "put down" after a few years?

> "Most health issues are caused by poor nutrition, unclean conditions, and over-crowding."

Most health issues goats face are caused by:

→ **Poor nutrition** (proper nutrition is vital to build and maintain a healthy immune system).

→ **Unclean conditions** (the housing, pasture and bedding MUST be kept clean and dry). Good stable as well as pasture management will go a long way to avoiding health problems in your herd.

→ **Over-crowding.** Disease spreads easily in crowded and unclean conditions. Provide adequate space for the herd.

BASIC REQUIREMENTS – MINERALS

Cows, sheep and goats often have a high need for mineral supplements. Regardless of where they are foraging, it is important to provide mineral supplements in the form of free choice blocks. This may provide them with the extras that they need to maintain the highest quality of health. Each species has different mineral requirements. Lumping them altogether with the same mineral needs may cause serious health problems.

Anemic - Iron: Needs to be given with copper to be assimilated and copper raises iron level quickly (kelp is a good source) or in an emergency an injection of iron squirted onto feed rations.

Bloat, nervous behavior or low milk production - Dolomite: Calcium (fluoride 45 ppm) and magnesium (20 ppm). Use in case of bloat by mixing with water and giving it as a drench – can be life-saving remedy.

> *"Free choice mineral blocks may provide them with the extras they need to maintain the highest quality of health."*

Creaking Joints - Boron: Borax (sodium borate) obtained through kelp (seaweed).

Dandruff - Iodine: Deficiency from overfeed, alfalfa and clover – obtain from kelp. Perhaps copper deficiency.

Failure to pass placenta, low birthweight, abortion - Selenium: A yellow sulfur (obtained from seaweed - kelp) that is often absent from soil in North American. If kelp is not available, jewel weed is a rich source (found in the eastern part of the U.S.). Need to feed with A&D oil if given as straight selenium (not kelp) to help absorption of the selenium.

Hair loss, skin irritation or doesn't conceive - A&D Oil: Vitamins are very important nutrients. All of the Vitamin B complex and Vitamin K are produced in the animal's rumen, and the body manufactures Vitamin C. That means that only Vitamins A, D and E are needed for balanced ruminant nutrition. Helps keep animal healthy during winter months when hours of daylight are short.

Immune System (poor) - Zinc: Obtained from kelp. Livestock need traces of zinc to maintain healthy immune systems.

> "Zinc can be obtained from kelp. Traces of zinc help maintain healthy immune systems."

Staph lesions on the body, a thin and faded hair coat, bald tail tips, twisting and bending of the front legs, spinal cord injuries or anemia, consistent worm problem, changing of hair color (a white goat with brown and black spots will become much whiter) or chewing wood - Copper: This important mineral can really influence the health of a livestock animal. Must be fed with iron (kelp is a good source) to be absorbed. Each animal species' requirement is different. For example, goats need up to 3,000 ppm (parts per million) and must have iron with the dose to help absorption into the system.

MINERAL DEFICIENCIES:

Mineral deficiencies in livestock diets can manifest themselves in a number of ways. These may include:

→ **Cobalt** (Vitamin B12) deficiency can result in a goat that displays: a loss of appetite, poor growth rate or weight loss, scaly ears, a watery discharge from the eyes, chewing wood, and/or bald tail tips. Young animals are more susceptible to this problem than adults. In some areas, cobalt deficiency is complicated by additional deficiencies in copper - and treatment is ineffective unless both copper and cobalt are provided.

→ **Copper**: Copper is essential for good pasture growth, as well as animal health. Copper is unique because of how it interacts with other essential elements. Changing of hair color (for example, a white goat with brown and black spots will become much whiter) is a sign of copper deficiency.

Excesses of both molybdenum (used in some commercial fertilizing sprays) and sulphur can cause copper deficiency in animals even if they are receiving the proper amount of copper.

Copper is more available to livestock from dry pastures and hay, so it tends to be a seasonal issue. Problems are worse in the winter-spring period and are often resolved over summer-autumn, even with no supplemental copper.

"Copper must be fed with iron and can really influence the health of a livestock animal."

Deficiency of copper has often occurred in conjunction with cobalt deficiency, and many affected areas have received copper fertilizer applications. Copper fertilizers appear to provide adequate pasture copper for many years (at least 15 years) and problems of copper deficiency in livestock in more recent times are usually introduced by excessive intakes of molybdenum, sulphur and possibly other elements such as iron. Molybdenum toxicity (molybdenosis, molybdenum-induced copper deficiency) may be due to naturally occurring molybdenum in pastures (for example peat swamps), can be a result of application of molybdenum fertilizers to pasture and occurs with the application of lime fertilizers to pasture to treat soil pH.

Copper can be quite toxic if given in too high of doses so extreme caution should be used. One tsp per month is the recommended dose. Copper can be top dusted on feed according to package directions (in tiny measures).

> *"Iodine deficiency is common in goats in the U.S."*

Copper toxicity can occur in 3 ways:
* Overdosing with copper supplements.
* Prolonged grazing on clover-dominant pastures.
* Prolonged grazing on heliotrope (contains a liver toxic alkaloid).

→ **Iodine** deficiency is present in the soil, and in the grasses and forage grown in some areas of the USA.

Therefore, iodine should be provided in stabilized salt. Iodine deficiency symptoms may develop with normal to marginal iodine intake in goats (so iodine should be provided in all cases). Symptoms include: enlarged thyroid; poor growth; small, weak kids at birth; and poor reproductive ability.

Iodine deficiency is common in goats in the US. Iodine is related to the functioning of the thyroid gland. The thyroid gland manufactures thyroxine. Thyroxine is required for the normal development of the fetus. Thyroxine does not pass from mother to fetus, so the fetus has to make its own. The iodine level of the doe during gestation is therefore very important. (Iodine does pass from doe to fetus across the placenta.) Regular dusting of kelp onto the food is a good source of iodine. Iodine deficiency makes newborn kids very susceptible to cold, wet weather and the death rate may be very high.

→ **Iron** deficiency is also seldom seen in mature grazing goats. Iron deficiency might be seen in young kids since there is a very low iron content in the dam's milk. This is more commonly seen in kids fed in complete confinement, and it shows up as anemia - gums are white and eyes are very white. Iron deficiency can be prevented by access to pasture or a good quality trace mineral salt containing iron.

→ **Magnesium** deficiency is less common in grazing goats than it is in cattle. Goats have a limited ability to compensate for low magnesium by decreasing the amount of magnesium they excrete. Both urinary excretion and milk production are reduced in a goat whose diet is deficient in magnesium.

→ **Phosphorus** deficiency results in slowed growth, unthrifty appearance, and occasionally a poor appetite. Birth defects and an uneven gait or bowed legs are also signs of deficiency. Phosphorus deficiency in grazing goats is more likely than a calcium deficiency.

→ **Salt** (NaCl) is a necessary dietary component in livestock health, but is often forgotten. If an animal licks the ground regularly then it is probably lacking salt. This does not present a nutritional problem, but may depress feed and water consumption in some areas where salt content of the drinking water is quite high. Salt blocks are often used to supply trace minerals.

→ **Zinc** deficiency may result in stiffness of joints, smaller testicles, and lowered sexual drive. Each animal species requires a minimal level of zinc in the diet (kelp is a natural source).

PRO-BIOTICS, NOT ANTI-BIOTICS

What should be done when rain makes the pasture a muddy swamp, or the stall doesn't get cleaned out as regularly as it should, or one of those nasty slugs slides across some deer poop and then gets gobbled up by a goat? No matter how clean or how attentive the goat herder is, at some point one of the animals is going to get sick.

Sound livestock isn't always possible...sometimes an animal is not all that it seems to be when it is purchased or when it's born. From the beginning days of an animal's life, or whenever they are feeling poorly, introduce **probiotics** (acidophilus and apple cider vinegar) into their system to give them every chance of building their internal immune system.

> *"Introduce probiotics into their system to give them every chance of building their internal immune system."*

This approach to wellness is the opposite of more traditional use of antibiotics...a killer of bacteria. Every mammal, goat, pig, cow or human needs a healthy gut.

Antibiotics work by killing all bacteria (good and bad). What is left is a blank slate, that is defenseless when bad bacteria is reintroduced into the animal's system. Gut health comes from a balance of bacteria (parasites included). Sickness and disease come from an imbalance of gut bacteria.

Probiotics give the animal lots of good bacteria to create an internal system that can usually handle small doses of bad bacteria or parasites. The good bacteria just gobbles up the bad stuff.

First Aid: The Great Eight

The Great Eight: Apple cider vinegar, diatomaceous earth, acidophilus, kelp, bee propolis, Spanish black radish, slippery elm, and massage – what do those eight things have in common? They are the basis for restoring or maintaining health in a goat. While it is important to look at the causes of the health challenge, the problem almost always comes back to an imbalance in the rumen caused by an imbalance in bacteria, parasites, food substances or injury.

Immune System Problems - Acidophilus is a probiotic, the opposite of anti-biotic. It can restore the bacteria balance in the gut so that the immune system can regain the ability to function properly. Acidophilus is also an excellent treatment for wounds and infections because it destroys bad bacteria, allowing the good bacteria to take control again.

> "The problem almost always comes back to an imbalance in the rumen caused by an imbalance of bacteria, parasites, food or injury."

Acidophilus (try Maxi Baby Dophilus) is available online or perhaps find it at the local health food store. If it isn't available there, then try finding it at the local feed store (TSC or Rural King) in the form of a product called "Probiotic". This product is primarily sugar (not good because stomach acid attacks the good bacteria, which happens a lot less if the bacteria is taken without food) with a little acidophilus thrown in. A higher dose than indicated on the package is usually needed to get good results.

Internal Healing - Spanish black radish powder should be added to the slippery elm mixture to add healing power after internal parasite problems. Black radish is very helpful in promoting internal healing after a bout of parasite over load. Add about 1/4 teaspoon to food rations.

The secret to successfully bringing back an animal from a severe case of parasite overload is the slippery elm and Spanish black radish powder mixture. Without adding the healing portion of the process to the parasite reduction process, the animal will remain internally scarred - and may still die from the injuries received due to parasite damage.

> "Give the message that you, the leader of the herd, are on the job."

Mineral Deficiencies - Kelp (dried seaweed) It is a nutrient-rich food that provides many different minerals, including iodine. Add a tiny bit of kelp to their feed or free feed. They will only eat kelp if they need it.

Mites, Lice, Some Parasites - Diatomaceous Earth adds 12 minerals to the diet but it also introduces a natural material, ground up diatoms, to the gut to cut and dry up parasites, or drive them out of the animal's digestive system. Can be dusted onto bedding or other areas used by livestock.

Poisoning - Livestock sometimes eat things that are poisonous. Remedy: mix 1 pint water, 1 tsp. ginger, ½ tsp. baking soda, ½ tsp. salt, 1 tsp. molasses, 1 tbs. Epsom salts. Give the entire mixture to an adult animal with an empty syringe.

Rumen quits working - Apple Cider Vinegar spritzed onto hay in the manger provides the animal a way to keep its rumen working properly by adding good bacteria to its feed. The vinegar encourages the rumen to keep moving and it also adds essential minerals to the diet. In a case where the rumen shuts down, the vinegar can be administered with a needleless syringe to the back of the throat for swallowing.

Shock or Bloat - Massage to the rumen, or an injured muscle, as part of the healing therapy plan, give the animal the message that you, the leader of the herd are on the job. Goats are herd animals and become nervous when there is no leader. A confident leader is reassuring to the animal.

In the case of a bloated rumen, the action of pushing and moving the rumen can assist it in doing its job when it is struggling from too much material inside or from too much acid generated when the rumen is not working. When massaging, use the tips of four fingers to push the rumen. Move clockwise over the entire rumen. Gradually work the finger tips around the side of the animal, continuing to push in circles until the entire rumen area has been deeply massaged.

Don't wait to practice this during an emergency - but use this technique while the animal is calm and able to accept touch. Otherwise, if she's in pain, and panicing, the stress level will increase and the massage will not be effective.

> *"Cover the wound with a mix of propolis, raw honey, acidophilus and black radish."*

Weak or Malnourished Animals – Nutri-Drench is a vitamin-rich liquid product that can be added to food or squirted directly in the mouth (sometimes used after childbirth for a weak animal) following illness or a bout with acidosis. Good for a burst of energy when the animal needs nutritional help to recover.

Wounds or Internal Healing (after worming) - Slippery elm powder has great healing power. Add one teaspoon to one tablespoon of bee propolis and one teaspoon of acidophilus powder and mix. Use on wounds or internally (after worming).

Wounds or Scrapes - Bee Propolis has antibacterial and anti-fungal properties that create "the inhibitory effect" (prevents infection while the immune system heals). Propolis contains many minerals and vitamins beneficial to all mammals and can absorb water from surrounding inflamed tissue promoting healing in ways that conventional treatments fail to do. For wounds or scrapes, cover the wound with a mix of propolis, raw honey, acidophilus and black radish.

SPECIALTY DAIRY FARMING RESOURCES: GOATS

FIRST AID

THE GREAT EIGHT:

Apple cider vinegar detoxifies the rumen, so do use this to help keep probiotics up. Spanish black (rather slippery elm, and minerals) is well-known to do eight things. Have a veterinarian. They are at the basics. The remaining is maintaining health is is good. which is important to look at the causes of the results of challenges, the problem is almost always comes back to an imbalance in the rumen caused by, an overuse of antibiotics, parasites, food contamination or injury.

EMERGENCY INFO

Poison Control Center: aspca.org | 1-800-222-1222
Animal Poison Control (aspca.org) | 1-888-426-4435

NAME
NUMBER
VETERINARIAN
EMERGENCY CONTACT

Immune System Resilience – Actophilus probiotic is the corporate immune system that keeps the battle. Its function properly. Acidophilus is part of an excellent immune process and when flora because of electrolytes and bacteria is lowered the goat becomes to have a complete.

Internal Parasites – Spanish black moss powder should be added to the always administered to be added including livestock or parasites, the water produces it has to also to use to help prevent that parasites by keeping the flora of the rumen in check and out of a raw area in the rumen.

The second to resting the body longer but shares them in a parasite of preventing is the slippery elm and diarrhea back rough powder makes. Without resisting the rumen is pulled of the process to the parasite reduction through the animal will remain chronically scoured, and more still die from its litter due to parasites.

Internal Desiccation – Keep blood renewed is a way is when bleed out of the feed can hardest. They are only a daily of the need of the animal's well.

Mites, Lice, Mange Parasites – Diatomeceous Earth acts as spearmint to the killing it, by absorbed and drying out of the animal. If the treatment will help to the animal is not used an alfalfa per bedding or other are used in a pasture and silken. Can be dusted into bedding of other animal near by livestock system.

Polaroid – Livestock some kind of things are not a really a remedy when a pair is also too carry of use. If only a small amount to take so so it is not to use it to have to carry any also used with the tape of the same out and to eat it ever.

Rumen quits working **Apple Cider Vinegar** sprayed on so feed is the original parasite remedy. Offering it is way of being eaten. Also stimulates well then on it with the rumen. The weight encourages the rumen to start moving and it also acids all ingest nutrients to the alfalfa of the rumen shots down, then they are on the administered with a copper syringe to the back of the throat of the throat for avoidance.

Shock or Injury – Massage to be a part of an open wound has to be worked on of the hell injury. This is well. In the case of a crisis in acupressure all point. In the case of a bleed it is to just be kept. Massage is a way to help the wound of it that by wed when it is changing from to you. In the main out as if it a have is it a drop keep when an eyes using down, from to stop working. Nothing will be with words the when you mean using over his for your true it means the put the rumen. As you massage down. You of over the entire rumen. Gradually, from the finger tips around the rub of the entire area, continuing to same circles, until it as it is that over stable to go recovered.

Do not try to use this doing on emergency, but as they learns to help make to call one be a put up of a panic to leave and like of a future get has it be a of trouble. And you can deal a with it be it to the one.

Nasal Diarrhea is a very much a liquid problem that can be added to biotin of goat and properly them. And with it and after using it in for a while, most that they leaves be a little of pouring about of the medication for a week meat to also be to look it tomorrow day others be bound of being before. Be. Use on record or intervals, on this done for it to be know.

Wounds or Abdominal Injury – Slippery skin covers area. The base of an active leaving it. The used the slimmy coat worn in to bandages for as much as but of wet it in for a with a wool on. The bandage the best but to stay for a use or contains.

Wounds or intestinal Hemitorg aches area – Bee Propolis excellent broad of an anti-fungal, antiviral tincture and area on the blood. The will only over the of to sticking. Can be admin the in one to cover the wound with a min of propolis one in to be of teaspoon to dilute.

Honey – a simple one food makin

Reference: Blue Book Edition

ESTIMATING WEIGHT

Knowing the normal weight of each goat in the herd is helpful in case of illness. Most feed stores have a special measuring tape for sale to "weigh" goats. The weight of a goat can also be estimated with the use of a simple measuring tape obtained from a fabric shop.

- → Measure the upper part of the goat's body just around and behind the hind legs (heart girth). (A)
- → Next, measure the length of the body from the point of the shoulder to the top point where the tail begins. (B)
- → Multiply A x A x B and divide by 300.

The resulting sum equals the approximate weight of the goat.

For example, a goat with a heart girth measurement of 38 inches, and a 30 inch measurement from the shoulder to pinbone (start of tail) = 38 x 38 x 30 / 300 = 144 pounds

For average size goats, the following chart can be used, however the estimated weights will not be as accurate.

Heart Girth (inches)	Weight (lbs)	Heart Girth (inches)	Weight (Lbs)
11 ¼	5 ½	27 ¼	69
12 ¼	6 ½	28 ¼	75
13 ¼	8	29 ¼	81
14 ¼	10	30 ¼	87
15 ¼	12	31 ¼	93
16 ¼	14	32 ¼	101
17 ¼	17	33 ¼	110
18 ¼	21	34 ¼	120
19 ¼	25	35 ¼	130
20 ¼	29	36 ¼	140
21 ¼	35	37 ¼	150
22 ¼	39	38 ¼	160
23 ¼	45	39 ¼	170
24 ¼	51	40 ¼	180
25 ¼	57	41 ¼	190
26 ¼	63	42 ¼	200

SUGGESTED RESOURCES

- → Horner, S. (n.d.). Calculating Goat Body Weights. Retrieved from http://www.infovets.com/books/smrm/C/C098.htm

Herbal Remedies: What They Do

(From https://libertyhomesteadfarm.com/herbal-remedies/)

- **Alfalfa** – Alfalfa contains large amounts of protein, minerals and vitamins; it is nervine and tonic and is an excellent kidney cleanser. Because alfalfa has roots that can go as deep as 125 ft., it brings up vital minerals not attainable by other vegetation. It is a rich source of vitamins A, C, E and K. It is a blood builder, good for teeth and bones, and excellent for milk producing livestock.
- **Birch** – Birch is useful in treating digestive ailments. The leaves are cleansing and can expel worms.
- **Carrots** – Carrots are useful for eye disorders due to the carotene. They are good for all livestock and help to expel worms.
- **Comfrey** – Comfrey is good orhealing bones, particularly the young shoots. Its healing substance is identified as allantoin.
- **Dandelion** – Dandelion is blood-cleansing and tonic and helps cure jaundice. The leaves strengthen tooth enamel and dandelion is an over-all good health conditioner.
- **Dill** – Dill increases milk yield and is a good treatment for digestive ailments.
- **Fennel** – Fennel increases milk yield and possesses antiseptic and tonic properties.
- **Garlic** – Garlic is very well known for its medicinal purposes. Highly antiseptic, it is rich in sulfur and volatile oils. Garlic is one of the best worm expellants and helps immunize against infectious diseases and assists in treating fever, gastric disorders, and rheumatism. It is also affective against parasites such as ticks, lice and liver fluke.
- **Hops** – Hop shoots are relished by grazing animals and are a good conditioner and/or tonic. It is also an antiseptic and vermifuge. Flowers are a milk stimulant.
- **Horehound** – Horehound is best known as a cough remedy in the treatment of pneumonia, colds, and lung disorders.

- **Lavender** – Lavender is highly tonic, antiseptic, antifungal, antibacterial, and gives a sweet flavor to milk and cheese. The whole plant is useful.
- **Lemon** – Lemon is a good blood cleanser. Also good for fevers, diarrhea and worms and may be used externally for skin ailments, ringworm and mange and to cleanse sores. Add honey when using internally.
- **Lemon Balm** – Lemon balm is a good pasture plant as it promotes the flow of milk. It's good for retained afterbirth and uterine disorders.
- **Marigold** – Marigold is eagerly eaten by livestock. It is a good heart medicine.
- **Mint** – Mint will decrease milk flow and would be good for livestock when weaning their young.
- **Mulberry** – Mulberry leaves and fruit are a good treatment for worms.
- **Mustard** – Mustard is a good natural dewormer
- **Parsley** – Parsley improves milk yield and livestock love it. It is rich in iron and copper and improves the blood. It also contains vitamins A and B and is good in cases of rheumatism, arthritis, emaciation, acidosis, and for diseases of the urinary tract.
- **Pumpkins** – Are excellent for deworming and are a good source of vitamins. Sheep particularly love pumpkins.
- **Raspberry** – Raspberry is well liked by goats. It is especially good for pregnancy and birthing. It is also good for digestive ailments.
- **Rosemary** – Livestock love rosemary and it gives a fine flavor to the milk. It is both tonic and antiseptic.
- **Sunflowers** – Sunflowers are rich in Vitamins B (1), A, D and E.
- **Thyme** – Thyme is another milk tonic and the oil is a worm expellant.
- **Turnips** – Turnips are another good food source that helps in deworming.

→ **Violet** – Violet leaves are rich in Vitamin C and A.

→ **Watercress** – Watercress has large quantities of vitamins A, B, C and B (2), as well as iron, copper, magnesium, and calcium. It promotes strong bones and teeth and is good for anemia. It increases milk yield.

→ **Wormwood** – This very powerful herb is especially good as a dewormer, as is Southernwood.

Disclaimer: These remedies are not intended to diagnose, treat, cure, mitigate or prevent any disease.

The information and statements presented here have not been evaluated or approved by the Food and Drug Administration (FDA). The use of herbs and essential oil for the prevention, treatment, mitigation or cure of disease has not been approved by the FDA or USDA. We therefore make no claims to this effect.

We are not veterinarians or doctors. The information here is based on the traditional and historic use of herbs as well as personal experience and is provided for general reference and educational purposes only. It is not intended to diagnose, prescribe or promote any direct or implied health claims. This information and product references are not intended to replace professional veterinary and/or medical advice. You should not use this information to diagnose or treat any health problems or illnesses without consulting your vet and/or doctor.

We present the products on this site and the information supplied here without guarantees, and we disclaim all liability in connection with the use of these products and/or information. Any person making the decision to act upon this information is responsible for investigating and understanding the effects of their own actions.

Parasite Control

On the surface of it, goats seem to have complicated digestive systems.

Each of the four goat rumens serve a purpose to aid in the health of the animal, and they all need healthy bacteria if the animal is going to maintain a healthy balance of bacteria.

It is also important to understand that parasites exist and that under a normal load (think balanced) in a healthy animal, there is no health problem. It is only when the parasite load becomes out of balance that bacteria and/or parasites adversely affect the health of the animal.

To maintain this healthy balance, several things are critical. These include:

→ Start with healthy stock (animals that have a rough start in life seem to have problems throughout their lives).
→ The right food, and clean water.
→ Clean and dry living conditions.
→ Proper population density (herd animals need company, but overcrowding is a common cause of many health problems).

"It is important to understand that parasites exist in a healthy animal."

There are many types of parasites, but for discussion/education purposes, the focus of this chapter is to outline the ones most encountered by goat herders.

Coccidia and Barber Pole worms have become the new "super bugs" in the goat and sheep world, and to make matters more challenging, these "bugs" have consistently managed to develop chemical resistance to all of the most commonly used products used to control the problem.

Because parasites have developed chemical resistance to most commercial products used to control or expell the worms/bugs it is now common practice to double the dose of the product, and pair it up with another product with different chemical components. *(See the chart at the end of this article for the proper dosages of these commerical products.)*

Many people who raise goats have moved beyond using chemicals for parasite control and are using medicinal herb mixtures to either kill the parasites or expel them (anthelmintic is an agent that destroy or expel parasites from the body).

> "Many people who raise goats have moved beyond using chemicals for parasite control."

Good sources for how to use herbal plant mixture remedies include:

→ Land of Havilah (https://landofhavilahfarm.com)
→ Katherine A. Drovdahl MH, CR, DipHlr, CEIT and her book *The Accessible Pet, Equine and Livestock Herbal*
→ *Holistic Goat Care* by Gianaclis Caldwell
→ Homemade herbal wormer formula from https://libertyhomesteadfarm.com/herbal-remedies/homemade-herbal-animal-dewormer-tonic/ is a pretty good one; add a little powdered lavender and slippery elm to it for healing.

High quality powdered herbs:
→ Mountain Rose Herbs (www.mountainroseherbs.com) is a reliable source for small quantities or to purchase in bulk.

Non-blood Sucking Parasites:
Weight loss, clumpy droppings, diarrhea, rough coat, and listlessness are all indicators that the parasite load is out of balance in a goat. If diarrhea

is involved in young goats, it's typically an indication of coccidia and treatment is necessary once the cause is confirmed.

In older goats, the symptoms of weight loss are usually an indication of non-blood sucking intestinal worms. A veterinarian or animal health care professional can analyze a fecal sample, identify the parasite, and recommend treatment. Or it's possible to use medicinal herbal remedies to return balance, but these require more time and effort because they focus on frequently administered daily doses.

Barberpole:
FAMACHA is a diagnostic test to help identify animals that require anthelmintic (an agent that destroys or expels parasites from the body) treatment and those which do not require deworming. The tool is a card that matches eyelid color to anemia levels, an indicator of clinical barber pole worm infection. Its use is limited to the parasites which cause anemia.

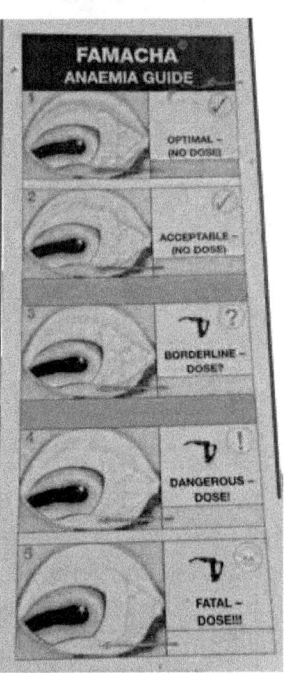

To use this system, the coloration of the goat's lower inner eyelid is compared to a FAMACHA color chart and those goats exhibiting coloration in the dangerous or fatal zone are treated with anthelmintics.

Another indicator of a dangerous level of blood-sucking worms is the development of bottle jaw. Any goat with bottle jaw (a large lump under the chin) should be immediately treated.

FAMACHA training classes are regularly held around the country and information on them can be found through the American Consortium for Small Ruminant Parasite Control website (https://www.wormx.info/resources).

Copper Oxide Wire Particles (COWP)

Low doses of copper oxide wire particles (COWP) have been shown to reduce barber pole worm infestations in goats. More recent research has demonstrated that COWPs may provide a natural method of controlling coccidiosis in kids. (American Consortium for Small Ruminant Parasite Control https://www.wormx.info/)

The tiny pieces of wire are available in capsules for easy administering in the right dosage. They can either by given with a bolus to get them down the goat or kid's throat or they can be top dressed on feed or worked into a peanut butter or a fig newton ball for goats that don't like or normally eat feed rations.

> *"As pigs root the soil, they are destoying the parasite life cycle, destroying eggs and young that have the potential to kill goats."*

Sericea lespedeza

Sericea lespedeza (Lespedeza cuneata L.) is a high-tannin forage that has been scientifically proven to reduce parasite loads in goats. More recent research has shown that sericea lespedeza pellets may offer natural control of coccidiosis.

Each goat has its own idea of what tastes good and they often don't initially like the sericea lespedeza pellets because they require more chewing then regular feed rations. It is possible to add the sericea in the form of seed to pastures to expand plant variety. For goats that don't like the taste, it is possible to grind up the pellets and top dress about a tablespoon on top of feed rations.

PIGS CAN CONTROL PARASITES:

Parasite management begins by working the soil, but not by cultivating or tilling it. Paddocks are sectioned off and pigs are rotated from paddock to paddock during the warm season of the year.

The job of the pigs is to do what they do best – root the soil. As they root, they are destroying the parasite life cycle, destroying eggs and young that are ready to begin a life that have the potential to kill goats. For best results, free-range chickens need to follow the pigs around, eating every little bit that the pigs miss.

One of the benefits of this type of management is that the pigs also increase the numbers of available plants within the pasture and increase the health of the existing plants. This makes for a pasture that is good for the goat herder and the goat herd. (Keep in mind though that some of the benefits from this type of plan may take years to realize.)

NOTE ON COMMERCIAL WORMERS

Commercial wormers are increasingly becoming ineffective against parasites. It is recommended to double the dosage on the wormers. Do not mix the wormers together – give them individually. Drench the wormer by tilting back the head of the animal slightly so that the substance goes to the back of the throat more easily. Give on an empty stomach to increase absorption in the gut.

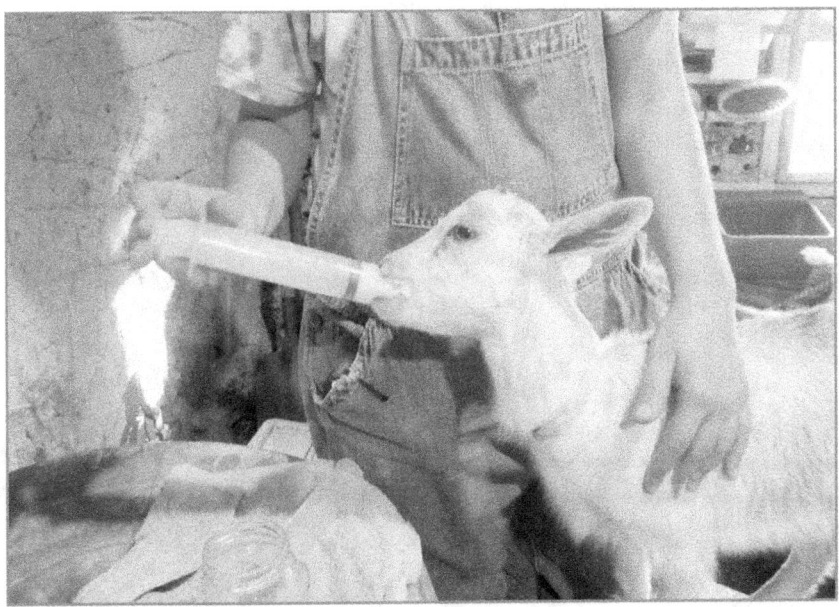

DEWORMER CHART FOR GOATS
RAY KAPLAN, DVM, PHD, UNIVERSITY OF GEORGIA

* IMPORTANT - PLEASE READ NOTES ATTACHED BEFORE USING THIS CHART *

1 ml = 1cc	Valbazen (albendazole) ORALLY	SafeGuard (fenbendazole) ORALLY	Ivomec Sheep Drench (ivermectin) ORALLY	Prohibit (levamisole) ORALLY	Cydectin Sheep Drench (moxidectin) ORALLY	Rumatel (morantel) Feed Pre-mix ORALLY
Weight Pounds (lbs)	20 mg/kg 2 ml/ 25 lb	10 mg/kg 1.1 ml/ 25 lb	0.4 mg/kg 6 ml/ 25 lb	12 mg/kg 2.7 ml/ 25 lb	0.4 mg/kg 4.5 ml/25 lb	10 mg/kg 45 gm/100 lb BW (Durvet)
20	1.6	0.9	4.8	2.2	3.6	
25	2	1.1	6.0	2.7	4.5	11 grams
30	2.4	1.4	7.2	3.3	5.4	
35	2.8	1.6	8.4	3.8	6.5	
40	3.2	1.8	9.6	4.4	7.3	
45	3.6	2.1	10.8	4.9	8.2	
50	4	2.3	12.0	5.5	9.0	23 grams
55	4.4	2.5	13.2	6.0	10.0	
60	4.8	2.7	14.4	6.6	11.0	
65	5.2	3.0	15.6	7.1	12.0	
70	5.6	3.2	16.8	7.7	12.7	
75	6	3.4	18.0	8.2	13.6	34 grams
80	6.4	3.6	19.2	8.8	14.6	
85	6.8	3.9	20.4	9.3	15.4	
90	7.2	4.1	21.6	9.9	16.4	
95	7.6	4.3	22.8	10.4	17.3	
100	8	4.6	24.0	11.0	18.0	45 grams
105	8.4	4.8	25.2	11.5	19.0	
110	8.8	5.0	26.4	12.1	20.0	
115	9.2	5.2	27.6	12.6	21.0	
120	9.6	5.5	28.8	13.2	22.0	
125	10	5.7	30.0	13.7	22.7	56 grams
130	10.4	5.9	31.2	14.3	23.6	
140	11.2	6.4	33.6	15.4	25.4	
150	12	6.8	36.0	16.5	27.3	68 grams

Herbal Wormer Recipe:

There are two ways to try to create a balance of internal parasites in a goat. Either kill the parasites with either a chemical or herbal dewormer product or change how the gut interacts with the parasite by making it uninhabitable by stunning or killing them but without causing significant harm to the animal. Changing the gut flora with this type of plant or drug is called "anthelmintics."

Herbs need to have other herbs to create a powerful formula for expelling or killing the parasites. A concoction of the following herbs have the capability, when used together to expel or kill parasites. Herbs can help expel the worms, and in some cases kill them.

It isn't enough to kill or expel the worms. Real gut healing needs to take place. Adding herbs like lavender, and herbal remedies like slippery elm powder and black Spanish radish helps the gut to be soothed and heal.

- **Cloves** are anthelmintic, coccidiostatic, and help firm up stool.
- **Anise** makes it more palatable (especially for dogs) and is antiparasitic.
- **Black walnut** hull powder is a strong anthelmintic**
- **Cayenne** supplies vitamin C and expels worms with the capsaicin.
- **Cinnamon** is coccidiostatic and antidiarrheal.
- **Garlic** is anthelmintic, antiparasitic, antibiotic, and antiviral***
- **Ginger** is astringent, anthelmintic, antibiotic, antidiarrheal, and coccidiostat***
- **Mustard** is antiparasitic and anthelmintic
- **Psyllium** expels worms by scrubbing the GI tract*
- **Rosemary** is antiparasitic and anthelmintic.
- **Sage** is anthelmintic.
- **Thyme** is anthelmintic, antiviral, antibiotic.
- **Wormwood** is a very strong anthelmintic**

By using a varied combination of anthelmintic (deworming) herbs, the herbal mix helps control a larger variety of internal parasites and support overall health.

*Psyllium husk powder may not be as effective as Psyllium seed powder. If the seed powder is available, that would be best, but the husk powder may help if it's the only thing available.

**Do not add Wormwood or Black Walnut for pregnant animals; do not add Black Walnut at all for equines. Dosage: 1 teaspoon for every 30 pounds animal weight, 1 tablespoon for every 100 lbs weight. (Use 1 cup of the powdered herbs to 2 cups of the cut herbs, except for the cloves, and diatomaceous earth)

***Garlic and Ginger: If the situation is acute (white eyelids, weak animal, off feed) I will add a 1/8th tsp. Of both garlic and ginger paste to the feed.

For the herbal de-wormer formula to work effectively it is necessary to give the right dosage, which is based on the weight of the animal. Recommended dosage: 1 teaspoon for every 30 pounds animal weight, 1 tablespoon for every 100 lbs weight.

(Note from Liberty Homestead Farm) It is better to overdose than under dose with herbs. Extra herbs will not hurt the animal, but too little will not be effective.

Pre-packaged goat de-wormers are quite pricey. While most of these herbs can be grown in the garden, Mountain Rose Herbs is a good source for bulk purchases. Diatomaceous earth is readily available in bulk from many sources.

This homemade version from https://libertyhomesteadfarm.com/herbal-remedies/homemade-herbal-animal-dewormer-tonic/ is a pretty good one, although adding a little powdered lavender and slippery elm to it for healing is even better.

HERBAL ANIMAL DE-WORMER INGREDIENTS:

- 1/2 cup whole or powdered Cloves
- 1 cup Anise Seed powder (optional)
- 1 cup Black Walnut hull powder*
- 1 cup Cayenne Pepper powder*
- 1 cup Cinnamon powder
- 1 cup Garlic (powder or minced)*
- 1 cup Ginger Root powder
- 1 cup Mustard seed powder
- 1 cup Psyllium seed powder
- 1 cup Rosemary leaf powder
- 2 cups Sage leaf
- 2 cups Thyme leaf
- 2 cups Wormwood*
- 2 cups Diatomaceous Earth —Note that diatomaceous earth makes the dewormer very dusty and hard on the lungs; it may be omitted and fed separately to promote the dewormer being palatable to the animal. Diatomaceous earth is very helpful in getting rid of external parasites as well. Use Diatomaceous Earth in the wormer for pigs, birds, ground feeds, and dosage balls.

The most crucial ingredients marked with an asterisk.

Directions:
→ Mix all ingredients together and store in a glass jar. Keep in a cool, dark place.
→ Administer for 7 days, morning and evening, for acute needs or as needed.
→ If Wormwood and Black Walnut have been left out of main batch:
   * Individual dosage of Wormwood is 1/4 tsp per 30# at same rate as dewormer (7 days, 2x per day, etc.)
   * Individual dosage of Black Walnut is 1/8 tsp per 30# at same rate as dewormer (7 days, 2x per day, etc.)

* Worming Kids: Administer the three-day dose starting at three weeks of age and repeating every 3-4 weeks until the kids are six months of age, then begin them on the eight-week cycle. If kids experience diarrhea (coccidiosis), immediately administer the concoction for three days in a row.

Herbal dosage balls can make herbs more palatable to livestock. Molasses is a good dosage ball base and it can be bought in bulk very cheaply at farm stores.

HERBAL DOSAGE BALLS FOR LIVESTOCK:

Mix together:
- → 1/2 cup (8 Tablespoons) powdered or finely crushed herb
- → 1/4 cup (4 Tablespoons) Slippery Elm Bark powder OR corn-based flour (this acts as a binding agent to hold the herb mixture together.)

Add:
- → 1/4 cup Molasses/Honey OR 1/2-2/3 cup Peanut Butter

Directions:
- → In a food processor (or by hand), mix and knead the mix into a dough.
- → Break into 12 even pieces, shape into balls and then roll the balls in a little bit of flour just to coat. Each ball equals a two (2) teaspoon dose.
- → Offer an herbal dosage ball to the animal first, and he may eat it readily. If he won't take the dosage ball, try introducing it by breaking off small pieces and placing it in the animal's mouth. (For goat kids, or other smaller animals, break the balls into smaller pieces to administer.)

WARNING: The rear teeth of livestock are EXTREMELY sharp and can give a very nasty cut. Be super careful during force feeding dosage balls!!

Pasture Management

What Goats Need

Goats are browser/foragers, not grazers. Plus they are herbivores, which means they are plant-eating animals. Goats use their lips and tongue to choose the tastiest plants.

As ruminants, goats swallow their partially chewed food into their first rumen, mix it there with some stomach acid, then regurgitate it back into their mouth (called "chewing their cud"), and then, swallow it again.

Goat metabolism makes them need to eat continuously, and they need a higher quality of plant material than is required by other ruminants in order to be healthy.

Goats are not grass eaters, although they will eat grass if that's all that is available. They do best by eating forbs (weeds) and browse (leaves of trees, shrubs, and vines that have woody stems).

"Goats need mineral supplements because the soil in North America is very depleted of these health-giving minerals."

Goats have a high need for mineral supplements because the soil in North America is very depleted of these health-giving minerals. Regardless of where goats are foraging, it is important to provide mineral supplements either as free choice blocks or loose minerals. This provides them with the extras that they need to maintain the highest quality of health. Requirements for minerals, trace minerals and vitamins are the same for meat and milk goats.

Goats need seven major minerals – calcium, phosphorus, magnesium sodium chlorine, potassium and sulfur. They also need many different trace elements such as selenium, copper and boron. Adding kelp to feed rations is a great one-size-fits-all way of bumping up the mineral rations. As a general rule, trace mineralized salt containing selenium should be given to all goats year around. Selenium is the one trace mineral that is often depleted in the soil in North America.

PASTURE RECIPE

The first thing to do to create any pasture plan is to conduct a comprehensive soil test. This can be done through companies like Crop Services (see Resource section for contact information) or Ohio Earth Foods.

> "Conduct a comprehensive soil test. The results will tell you everything you want to know about the soil condition."

The process is simple. Use a zip lock plastic bag for the soil contents. Take a small amount of soil from several areas to be tested so that there are about two (2) cups of soil. Fill out the online form for the sample, and then put it in the mail. The results from the test will outline the soil composition, including mineral content.

Minerals are super important in the recipe for creating healthy soil and livestock. Much of the pasture land in North America is depleted of the trace minerals necessary to grow plants and provide microbes needed for optimum goat health.

According to the experts, 90% of soil function is transformed by microbe organic matter providing plant nutrients that are assimilated by plants. Microbes depend on plants for their survival. For plant life to provide the highest nutrition to livestock it is vital to focus on the type, the variety and how to attract those microbes. Good pasture health can save a lot of money, time and heartache.

Pasture Requirements

Pastures need a few ingredients to get the formula right for attracting microbes, and feeding the plants that livestock will need to stay healthy.

→ A variety of plant types to create strong root systems, lots of those amazing microbes, plus attract beneficial insects.
→ Rotational grazing that offers nutritious mineral-rich food for livestock, and also lets the pasture rest.
→ Pasture recovery period – the longer the rest period, the more productive the pasture.
→ Smaller paddocks for rotating livestock so that the pasture isn't over-grazed. The more paddocks, the better.
→ Pasture rotation of animals when there is no less then 50% of the grasses at a minimum of 6" to 8" high. This will give the root system energy for re-growth through photosynthesis (the ability to produce sugar that spurs growth). Grazing at 10% more will reduce root growth by 50%, delaying the uptake of growth in the plant.

> *"Creating a plan that allows the pasture to function year round will keep the soil, plants and livestock healthier."*

→ Creating a plan that allows the pasture to function year round will keep the soil, plants and livestock healthier. When pasture is grazed to the soil level, it heats up more quickly and looses moisture. If the ground temperature is too warm livestock will avoid eating, and move to cooler locations. If the ground temperature is cooler because of plant cover, there will be less over fertilizing in cooler areas, and a better distribution of manure.
→ Know the condition of your pasture? Walk the pastures with a measuring stick to determine plant height, and the amount of dry matter available in the field.
→ Develop the ability to judge browser growth weeks in advance to determine how best to rotate the pastures. Depending on the current condition of the pasture, and the forecast for the next 10 days,

it might be necessary to rotate animals more frequently to ensure that the pasture has the time needed to recover. Too much rain, or too little can make a huge difference in how pasture plants recover from grazing.

→ The time for resting the paddock may be as short as seven (7) to 10 days in early spring, and by mid-summer perhaps 45 days. To know how to plan the rotational grazing requires analyzing how the plants are growing, moisture levels that the plants can draw on, and normal growing conditions for that month of the year.

> *"In order for photosynthesis to occur, plants need water, carbon dioxide, chlorophyll and light."*

According to John Kempf, founder of Advancing Eco Agriculture, "There is tremendous untapped potential that exists in forage (browser) productions."

In order for photosynthesis to occur, plants need four things - water, carbon dioxide, chlorophyll and light.

→ Carbon dioxide is naturally released from organic matter (think manure, deep bedding from the barn).

→ Water comes from rain, but during drier spells if there was more moisture available to the plants then they would continue to grow.

→ Chlorophyll: Nitrogen, magnesium, manganese, phosphorus, and iron are nutrients that give plants their dark green color. Iron is the pipe that pulls chlorophyll together.

→ Manganese support water hydrolysis – the process of how the plant spits water molecules to make chlorophyll.

→ Phosphorus aids the plant in converting the sugars it makes into energy so that the plant can grow.

If the plant is going to make more sugar, it needs more energy (made from the sugars the minerals and water provide to the plant) to assist in converting carbohydrates into biomass.

By late summer, the nutrients needed for the plant to grow are used up by the development of the flower and seed heads (the best food for goats). As the seed heads develop, the root system runs out of its energy supply, and the entire plant is vulnerable to heat, lack of moisture, and low carbon dioxide levels.

→ If plants are eaten by livestock (or clipped off), the roots keep supplying energy for plant growth instead of seed production.

→ Clipping too much or too often significantly reduces the energy in the root system, slowing leaf growth.

→ The frequency of clipping or grazing depends on many factors including the time of year, temperature, moisture, and soil fertility.

> *"Fences can be death traps for goats and their kids."*

Suggested Resources

→ McDonald, P., Edwards, R.A., Greenhalgh, J.F.D., Morgan, C.A., Sinclair, L.A & Wilkinson, R.G. (2011) Animal Nutrition. Seventh edition. Pearson

Fencing for Goats

The type of fencing used is important.
There are two types of fencing that work best for goats: polywire polytape and high tensile.

Fences can be death traps for goats and their kids. If the wrong polywire fence is used they can put their head through reaching for plants on the other side of the fence, and then cannot figure out how to get back through the fence.

Other goats can head butt them out of orneriness or from wanting whatever plant that's on the other side of the fence. If they're on a hillside they can lose their footing and hang themselves.

Temporary electric fencing (Premier1.com) is useful for easily changing the location and size of the paddock, but without training the goat on how to respect the fence, there can be serious problems, even death.

Temporary Wire Fencing
Polywire and polytape are combinations of braided, UV-stabilized polyethylene plastic interwoven with three to nine stainless steel, copper or aluminum filaments. Polytape is similar in composition to polywire but is flat, five-eighths to 1.5 inches wide and is used because of its visibility. Polywire fencing is more expensive and permanent then polytape.

"Electric fences require that goats go to fence kindergarten."

Polytape and polywire fencing can be used very effectively for grazing situations. Electric netting, a prefabricated fence of electro-plastic twines and white push-in insulated plastic posts can be used with goats but for these types of fences to be safe and effective it is necessary to train the goat over a period of time on how to use the fence. Goat herders have lost animals that tangled their horns in the netting, shocking them to death or strangling them.

Electric fences should be charged at 4,500 to 9,000 volts at all times. Lower voltage does not get enough of a shock to prevent the livestock from challenging the fence.

Regular checking and testing of the fence is necessary, and any problems must be fixed promptly, or goats will escape, or get caught in the webbing.

Electric fences require that goats go to fence kindergarten.

The best way to educate the goat about how the fence works is to use a small portion of the fence to create a circle big enough to easily hold several goats in a way that allows them to move around with lots of options for browsing. Make sure, if it's a plant-only paddock that the plants are tall and look inviting to the animal.

The goat should be "introduced" to the fence by leading it to the mesh netting (which is electrically charged) and allow the goat to touch the fence. The animal will receive a sharp shock. The fence is designed to shock and release, then shock and release.

Take the animal around the fence demonstrating how the fence works in several locations. The best results are often achieved because the animal will loose interest in what's just happened and go back to browsing.

No goat should be left alone in the fence during the kindergarten phase, especially if it has horns, until it's clear that it understands to avoid the mesh fence. During the kindergarten period it's vital to not leave the animals inside the fence for more then an hour.

> *"As a rule, goats will crawl under rather than jump over a fence."*

The next day put the animals inside the fence and stay with them until it's clear they will respect the fence, but don't leave them there for more then two hours this time. Continue this process until you are confident that they know to respect the fence. Now they are ready to stay inside of the fence during the day time. They should not be left there overnight because there are too many possible dangers with the fence to leave them in a temporary fencing situation.

Permanent Fencing

Permanent fencing for goats calls for 12.5 gauge, smooth, high-tensile, class 3, galvanized steel wire. Perimeter fence height should be at least

42' tall. A high wire (electrified), or an offset wire set one foot inside the fence near the top, may be needed if goat jumping is a problem.

As a rule, goats will crawl under rather than jump a fence, so the bottom wire should be kept close to the ground. Boundary fences should control all stock at all times. However, interior and cross fences may be made of three to four smooth strands of high-tensile wires, assuming animals are well trained.

Woven wire -- 6 inch-by-6 inch, or 6 inch-by-9 inch, openings -- is very effective as a permanent fence, but costs at least twice that of a five strands of smooth high tensile electric fence. Further, horned goats frequently become caught in the 6 inch-by-6 inch openings openings split by a T-post.

> *"Because goats like to climb, the corners of fences should not have diagonal bracing for posts."*

To address this problem with existing fences, an electric wire offset about 9 inches from the woven wire fence and about 12 inches to 15 inches from the ground will reduce the number of animals caught in the woven wire fence. However, this practice also reduces control of forage growth on the fence line.

Woven wire with a 6-by-9 inch opening is a new and cheaper alternative than the woven wire with a 6 inch-by-6 inch opening, which does not require an electric offset wire. A new 12.5 gauge high-tensile woven wire that has a 24 inch vertical spacing is now available. It is 36 inches tall and can have either five, seven, nine or eleven horizontal wires. It is generally less expensive than conventional woven wire.

Horned goats usually do not get caught in the woven wire fences with vertical spacing greater than 6 inches, or, if caught, they are able to free themselves because of the larger openings.

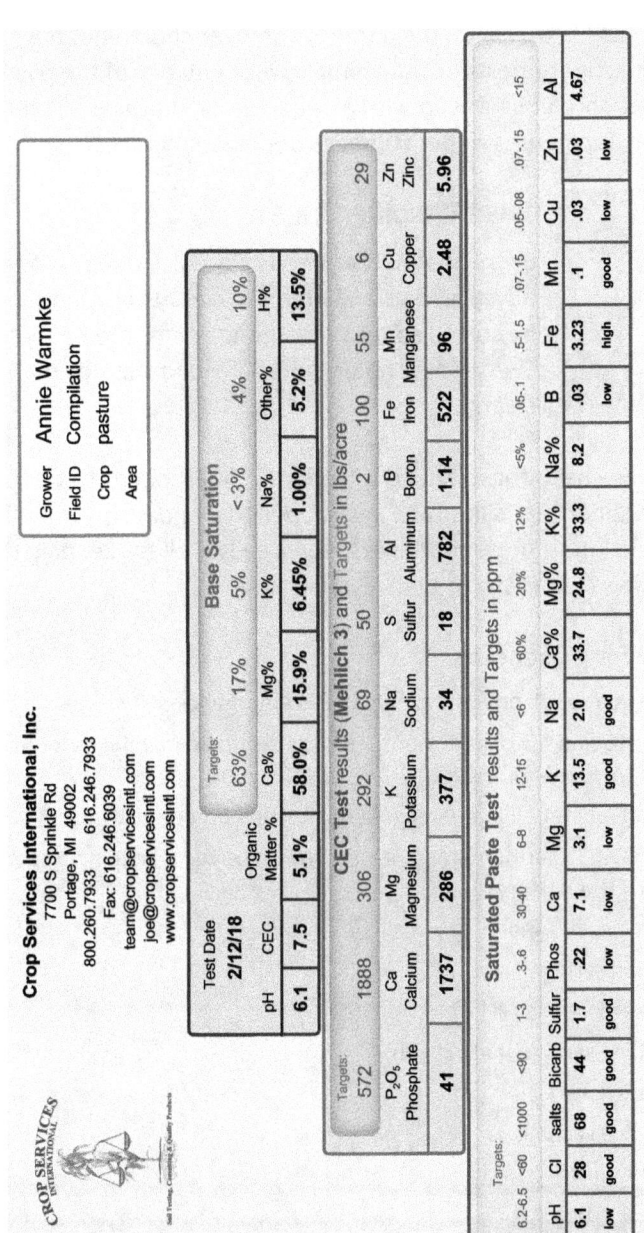

Crop Services International, Inc.
7700 S Sprinkle Rd
Portage, MI 49002
800.260.7933   616.246.7933
Fax: 616.246.6039
team@cropservicesintl.com
joe@cropservicesintl.com
www.cropservicesintl.com

Grower: Annie Warmke
Field ID: Compilation
Crop: pasture
Area:

Test Date: 2/12/18

pH	CEC	Organic Matter %		Base Saturation					
			Targets:	63%	17%	5%	<3%	4%	10%
				Ca%	Mg%	K%	Na%	Other%	H%
6.1	7.5	5.1%		58.0%	15.9%	6.45%	1.00%	5.2%	13.5%

CEC Test results (Mehlich 3) and Targets in lbs/acre

Targets:

572	1888	306	292	69	50	2	100	55	6	29
P₂O₅ Phosphate	Ca Calcium	Mg Magnesium	K Potassium	Na Sodium	S Sulfur	Al Aluminum	B Boron	Fe Iron	Mn Manganese	Cu Copper	Zn Zinc
41	1737	286	377	34	18	782	1.14	522	96	2.48	5.96

Saturated Paste Test results and Targets in ppm

Targets:

6.2-6.5	<60	<1000	<90	1-3	.3-.6	30-40	6-8	12-15	<6	60%	20%	12%	<5%	.05-.1	.5-1.5	.07-.15	.05-.08	.07-.15	<15
pH	Cl	salts	Bicarb	Sulfur	Phos	Ca	Mg	K	Na	Ca%	Mg%	K%	Na%	B	Fe	Mn	Cu	Zn	Al
6.1	28	68	44	1.7	.22	7.1	3.1	13.5	2.0	33.7	24.8	33.3	8.2	.03	3.23	.1	.03	.03	4.67
low	good	good	good	good	low	low	low	good	good					low	high	good	low	low	

Because goats like to climb, the corners of fences should not have the diagonal bracing for posts or the animals will climb out of the pasture. Corner posts should be driven with a deadman or H-braces. High-tensile electric fences can uses either H-braces or diagonal braces.

The Medicine Chest

Every goat herder needs a medicine chest for everyday remedies and emergencies. Everything should be sorted so that emergency items are easily found. Every item should be easily identified or marked clearly.

The medicine chest doesn't need to be fancy, but it needs to be some type of container that cannot be easily opened by goats or other livestock. Plastic multi-draw containers work really well for keeping things sorted, and as clean as possible.

A Checklist for most needs:
- → A good goat book on natural health and emergencies
- → An easy to find list of goat websites and Facebook goat pages for information and/or emergencies
- → The SARE posters for general goat health, and emergency information
- → Rectal digital thermometer (they are inexpensive so keeping a spare is helpful if the battery dies)
- → 3 cc, 6 cc, 12 cc and 35 cc plastic syringes
- → Weak kid syringe
- → Needles (1" and 1/2")
- → Drenching syringe (balling gun)
- → Surgical scissors
- → Disposable scalpels
- → Hoof trimmers
- → Brush for grooming
- → Spray bottle (for apple cider vinegar for cleaning feed pans and hands)
- → Small scoops and containers for dispensing powders

- → Extra small plastic bags and bottles for holding items that need to stay clean and organized
- → Calendar for keeping track of mating, birth dates and other important items
- → Small notebook in a zip lock bag for keeping notes
- → A 3-ring notebook for keeping medical records of each animal
- → Surgical gloves (with hair elastic bands for keeping them tight on the wrist so they don't slip off)
- → Blood stop powder or cayenne powder
- → Propylene glycol (for ketosis)
- → Probiotic powder/paste
- → Acidophilus powder for probiotic dusting on kid umbilical cord
- → Ketocheck (to diagnose ketosis)
- → Items to make electrolytes
  - molasses
  - salt
  - baking powder
  - apple cider vinegar
- → Bee propolis with raw honey
- → Slippery elm powder
- → GI Soother (for healing digestive system after diarrhea or parasite imbalance)
- → Charcoal (for bloat or diarrhea)
- → Vet wraps (for breaks or sprains)
- → Comfrey tincture (for bone breaks)
- → CBD Oil (for shock, pain)
- → Dr. Bach Remedies (for nervous system balance in case of shock or injury, transport, or pregnancy labor)
  - Rescue Remedy
  - Larch oil
  - Walnut oil
- → Garlic (pressed garlic in a jar of water for bumping up the effectiveness of parasite remedies/meds)
- → Tape measure for checking goat weights

GOAT ANATOMY & GOAT RUMEN

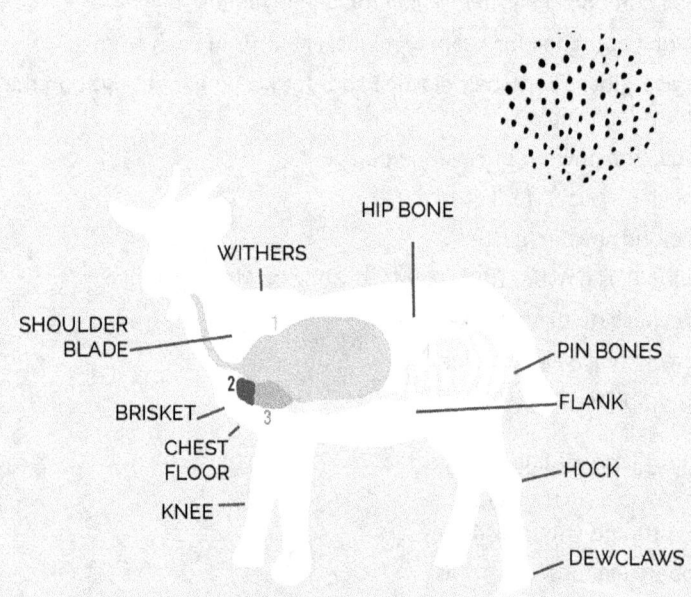

1. RUMEN 2. RECTICULUM 3. OMASUM 4. ABOMASUM

Woman's Story: Carie Starr
Carie Starr, Co-owner of Cherokee Valley Bison Ranch

I grew up on the farm where I now operate, Cherokee Valley Bison Ranch. The ranch originally belonged to my grandmother, who raised her children to respect the land and never allowed the use of any herbicides or pesticides on the farm. This respect of nature really made an impact on me. My grandfather always treated me like one of the boys, so I didn't learn that girls can't do certain things just because they are girls. That might have made all of the difference in how my life led me to be a bison rancher.

After I studied environmental science at Zane State College, I worked in several water quality labs, but the urge to farm was always calling to me.

My life was changed forever the first time I tasted bison at Ted's Montana Grill in Columbus, Ohio. I thought it was the most delicious food I ever ate. And shortly after that I realized that my 25-acre ranch might just support a herd of bison.

In 2008, my husband Jarod and I bought a herd of bison. By that time I had been studying about bison for a couple of years and decided to start small with just a few calves.

One Sunday Jarod saw an ad in the paper for a herd of bison. As it turned out, the farmer was going to send the herd to the processor if no one bought it. When I saw Charlie, the most picture-perfect bison I'd ever seen, it was love at first sight. We bought the herd of 14 bison on the spot.

It didn't take long to figure out that we didn't have adequate infrastructure - no barn, fence, or even a truck capable of hauling these large animals. Within six weeks we had built strong fences and had the bison delivered.

Becoming a full-time farmer takes time and money. So I worked my other job at first. Eventually we created a bison burger business, working the fair and festival circuit. However, it soon became clear that this approach would not work long term due to established competition and the high entrance and profit-sharing fees. So we bought some dairy goats to help fund the bison project. We scraped along by selling the baby goats and hosting a farm camp.

> "The problem was that they were trying to keep goats with bison fencing, so they got out of goat herding."

New to goat herding as well, I soon figured out this wasn't going to work either. I bought cheap stock, which led to lots of problems. I spent nearly half my day chasing goats, which helped develop my cussing vocabulary - but not a lot of revenue. Another problem was we were trying to keep goats with bison fencing - so we got out of goat herding.

The best advice I ever received was, "Don't be afraid to try." I tried keeping goats, and it just didn't work out, but we tried.

You can avoid lots of problems by doing research before jumping into raising livestock.

I will be the first to admit that I made a lot of mistakes along the way, but constantly evaluating what works and what does not - not being afraid to try new things or to abandon that "great idea" when it fails - has eventually led to a successful and enjoyable lifestyle.

WOMEN GROW OHIO - WOMEN MENTORING WOMEN

Women Grow Ohio connects and unites Ohio women in all forms of agriculture, including: livestock farmers, growers, homesteaders and urban backyard garden producers. Women in farming are scientists, economists, foresters, veterinarians, and conservationists.

Women are in the boardrooms and the corner offices of international enterprises, and are the owners and operators of small businesses. Women are property owners and managers. Women are policymakers and standard bearers. Women are involved in every aspect of agriculture. From historic homesteaders to contemporary cattle ranchers, women have been the cornerstone of Ohio's agriculture heritage.

"Mission: to connect and unite Ohio women in all forms of agriculture."

Women Grow Ohio works to provide networking opportunities, encouraging women to promote each other, offering peer-to-peer education and encouraging sustainability not only in agricultural activities, but in all aspects of women's lives.

Why are these things important?

A woman in ANY predominantly male-oriented vocation can find it lonely and intimidating. It is particularly so for a beginning woman farmer.

When most of the women you know are interested in jewelry parties and shopping and you just want to talk about chickens and why your peas didn't germinate, you begin to feel like you can't possibly fit in. There is a constant sense of isolation.

Male counterparts repeatedly demonstrate that we, as women farmers, simply aren't taken seriously.

> "Women need a voice in decisions and a safe place to ask questions and learn in a way that creates confidence and a sense of being valued."

"Oh you have five acres, aren't you cute. Where's your combine?"

"Hey, do you want me to back your stock trailer in there for you?"

"Do you want me to put that ratchet strap on for you?"

While many are simply attempts at being gallant, these statements and gestures can make women feel inept and dismissed as being "legitimate" farmers.

There is a very definite lack of women mentors to show the way in this male-dominated field.

If a women didn't grow up on a farm it can be intimidating at every level. But even for those who were raised on a farm, the rules have changed in the farming world and learning from other successful women is vital to achieving the ability to earn a living.

Women need a voice in decisions, and a safe place to ask questions and learn in a way that creates confidence and a sense of being valued.

Get in touch with Women Grow Ohio (www.ruralaction.org) to learn more about mentoring opporunities.

Woman's Story: Celeste Taylor
Farmer at Black Sheep Farm and Goat Herd Manager at Integration Acres in Athens, OH

My job at Integration Acres involves making sure all the goats are in good health.

During the off season, when the goats are not lactating, I only need to go to the farm every three days since one of my co-workers works on the farm full time. One of my jobs is to give the goats a small ration of grain to make sure they're keeping good body mass over the winter.

Goats are very good at saying, "You're at the bottom of the totem pole, so no food for you." So I need to ensure all the goats get their ration of grain.

Another role is to make sure that nothing has happened to the goats. They can often get gored or injured, especially when they are pinned up in a small paddock during the colder months.

In the spring, when the goats start kidding, my job is to make sure all the kids come into the world alive. Once the babies are born, I help the farm owners to figure out which kids should be sold and which ones will be raised to become milk goats.

A farm manager has to balance the number of animals with the amount of space available. I organize rotational grazing to help manage the land. This ensures the paddock growth is keeping up with the animals and the animals are keeping up with the paddock growth.

I also milk goats. At any given time, the farm may have four to five goats that need to be milked. Milking shifts are split between the farm staff to ensure the goats are well cared for and that the dairy is clean and well maintained.

Sometimes I help with the cheese making and clean the milk tanks as well as any other items associated with milking. I enjoy doing many of the mechanical repairs necessary to keep the operation functional.

I believe that, "Where there is livestock, there is dead stock." While it is nice to talk about cute babies, it is important to keep in mind that every single one of them will die eventually - whether they die young from some accident, or they die of old age or are butchered. It is often hard to deal with death, especially if the farmer gets emotionally attached to their animals.

"Be prepared for self-recrimination and the emotional roller coaster of living with death."

Most good farmers tend to blame themselves when things go wrong. Sometimes there is simply nothing that can be done to prevent a death. An animal can get its head stuck in a fence or feeder, or be injured by another goat - nothing can be done - but it doesn't make it less painful.

It is the stupid, preventable death, that is the hardest. Or the deaths where you try and try and try to save the animal but it still dies.. Be prepared for self-recrimination and the emotional roller coaster of living with death. Personally I think ice cream helps.

I keep sheep and goats on my own farm. I practice rotational grazing to assist in revitalizing the coal stripmined damaged land that I farm.

Another big consideration in using sheep for a dairy is their genetic make-up. Sheep have primarily been bred for meat and fleece, and are not typically bred for dairy. Having a sheep breed that is well suited for milking is critical; otherwise, a farmer could have to milk their herd as much as eight times a day.

Raising sheep requires a hybrid of goat and cow infrastructure. For example, a goat needs really aggressive fencing, because they can work through nearly any fence; whereas sheep have a very high prey drive and when they are frightened they will slam their bodies against whatever is in their way. Sheep fence must be heavy duty enough to hold a cow and tight enough to hold a goat.

> "Sheep, in particular, are more motivated by fear than by food."

Sheep in particular are more motivated by fear than by food, and they will react in wild and crazy ways. It is traumatic for them to be milked. So starting with baby sheep, handle the sheep often and aggressively to train them to be milked. Also make sure the milking infrastructure has lots of rounded edges so that the sheep cannot hurt itself.

Farmers should remember that temperament is something one can breed for. Choosing a calm mother can help ensure calm babies, and eventually a calm herd.

SPECIALTY DAIRY FARMING RESOURCES FOR GOATS

RECOGNIZING A SICK GOAT
When an animal is not feeling well, they may:
- Stop eating or have little interest in food.
- Stop drinking water or have little interest in water.
- Have diarrhea or clumpy stools.
- Stop urinating or is urinating painfully.
- Seem depressed – stands off alone from the herd or in the corner.
- Stop chewing her "cud".
- Seem to have bloated – has a larger than normal stomach
- Try to lay down, get up and cry and then try to lay down again

There are several things to look for in order to identify some sources of infection.

1) Look for any external cuts or swelling. A cut or scratch may not be obvious.
2) Look for leg limping or tenderness. They may limp because of an internal infection of a joint which can cause the leg to be extremely tender. Pregnancy toxemia will also cause an animal to limp but this is not an infection.
3) Look for signs of respiratory problems. Colds or pneumonia will cause a high temperature.
4) Take their temperature. If they have a temperature this may help to figure out where to start with a diagnosis.
   Reference: Blue Rock Station

APPROXIMATE MEASURES
LIQUIDS: (Please note: 1ml is the same as 1cc)
- 20 drops = 1 ml = 1cc
- 1 teaspoon = 5mls = 60grains = 60drops = 5grams
- 1 tablespoon = 3tsp = 15mls = 1/2ounce = 15grams
- 1 ounce = 30mls
- 1 cup = 16TBSP = 1/2pint = 8 fluid oz = 250mls
- 1 pint = 2cups = 16oz = 500mls
- 1 liter = 1,000mls

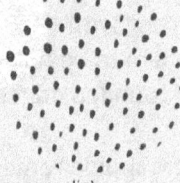

SOLIDS: (Please note: 1 gram weighs about the same as one regular paper clip)
- 1 mg (milligram) = 1/65 grain (gr)
- 1 mg = 1000 mcg (micrograms)
- 1,000 mg = 1 gram (g)
- 1 gram (g) = 15.43 grains (15gr).
- 1,000 g = 1 kilogram (kg) = 2.2 pounds (lb)
- 1 grain (gr) = 0.065 gram = 60 mg
- 1 ounce = 1/16 pound = 28.4 grams
- 1 pound = 0.454 kg = 454 grams

GESTATION LINK: http://americangoatsociety.com/ education/gestation_calculator.php

FEEDING

Dairy goats need a year-round supply of roughage, such as pasture, browse or well-cured hay. Winter browse and pastures should be supplemented with hay. Milking, breeding and growing stock need a daily portion of legume hay, such as alfalfa. Kids and bucks need a balanced grain ration and milkers should be fed a standard dairy grain ration. Kids are milk fed until two to three months of age, but should be consuming forages such as pasture grass or hay by two weeks of age and grain within four. All dairy goats must have salt and fresh clean water. Mineral supplements are desirable.

Reference: American Dairy Goat Association

PHYSIOLOGICAL VALUES OF GOATS

Pulse: about 83 per minute ranging from 50 to 115
Respiration: about 29 per minute ranging from 15 to 50
Body Temeperature: about 103.6 degrees Fahrenheit
Reference: American Dairy Goat Association

FACEBOOK RESOURCES

Totally Natural Goats & More!
Successful Goating with Rosie
Goat Emergency Help and General Questions

EMERGENCY INFO

NAME : _____

PHONE NUMBER : _____

VETERINARIAN : _____

EMERGENCY INFO

Poison Control Center (aapcc.org) 1-800-222-1222
Animal Poison Control (aspca.org) 1-888-426-4435

FIRST AID - THE GREAT EIGHT

Apple cider vinegar, diatomaceous earth, acidophilus, kelp, bee propolis, Spanish black radish, slippery elm, and massage – what do those eight things have in common? They are the basis for restoring or maintaining health in a goat. While it is important to look at the causes of the health challenge, the problem almost always comes back to an imbalance in the rumen caused by an imbalance in bacteria, parasites, food substances or injury.

- IMMUNE SYSTEM PROBLEMS - Acidophilus
- INTERNAL HEALING - Spanish black radish powder
- MINERAL DEFICIENCIES - Kelp (dried seaweed)
- MITES, LICE, SOME PARASITES - Diatomaceous Earth
- POISONING - Livestock
- RUMEN QUITS WORKING - Apple Cider Vinegar
- SHOCK OR BLOAT - Massage
- WEAK OR MALNOURISHED ANIMALS – Nutri-Drench
- WOUNDS OR INTERNAL HEALTING (after worming) - Slippery elm
- WOUNDS OR SCRAPES - Bee Propolis

Reference: Blue Rock Station

Podcasts and Webinars

SARE Specialty Dairy Webinar Schedule (available on ACEnet's YouTube channel, WGO website and at www.bluerockstation.com)

- → **Setting the Course: Making a Living with Specialty Dairy Products in Appalachian Ohio,** Presenters: Annie Warmke
- → **An interview with Sasha Sagetic** - Black Locust Livestock & Herbal: How she became a farmer and her philosophy.
- → **We 'Goat' Your Back, Dairy Do's and Don'ts from Veteran Dairy Goat Operators,** Presenters: Annie Warmke
- → **An interview with Carie Starr** - Cherokee Valley Bison Ranch: Her experiences in goat herding and the reasons she stopped.
- → **Creating Legen-'dairy' Products: Current Trends and Market Opportunities for the Specialty Dairy Industry,** Presenters: Leslie Schaller
- → **An interview with Annie Warmke** - Blue Rock Station: Annie talks with Catlyn Harrier about how she became a goat herder and her experiences with goats.
- → **Storytelling for Success: Marketing Your Business,** Presenters: Leslie Schaller, Sarah Cornwell
- → **An interview with Leslie Schaller** - ACEnet: She discusses the role of ACEnet and her history with local foods.
- → **Getting Market Ready: Pricing Specialty Dairy Products,** Presenters: Leslie Schaller, Sarah Cornwell
- → **An interview with Michelle Gorman** - Integration Acres: Discusses her business of cheese making and the challenges.
- → **Ready for Retail Sale,** Presenters: Leslie Schaller, Madelyn Brewer
- → **An interview with Abby Turner** - Lucky Penny Farm: How she came to be a goat herder and how she started her creamery
- → **Scaling up for Wholesale,** Presenters: Leslie Schaller, Madelyn Brewer
- → **An interview with Celeste Taylor** - herd manager at Integration Acres: She raises sheep and discusses sheep herding and dairy goats
- → **Budgeting: How Can I Make a Living Doing What I Love?** Presenters: Annie Warmke, Carie Starr
- → **Podcast: Dr. Rachel Terman** - Ohio University professor: Discusses with Annie Warmke general issues related to women and farming.
- → **Podcast: Becky Rondy** - Green Edge Gardens: Discussed her business model and the reasons farmers need to mentor and pass their knowledge on the new farmers.
- → **Mentoring & Networking with Women Grow Ohio,** Presenters: Annie Warmke, Carie Starr

Resources

Wellness

- CFAMACHA and body scoring charts: https://www.bing.com/images/search?view=detailV2&ccid=3Sv6Mjw%2f&id=BACBCC7F0E30AC5CD332BA28C8D9E11CFDC785CA&thid=OIP.3Sv6Mjw_BuWcN-ywQAqNjwHaFj&mediaurl=https%3a%2f%2fimage.slidesharecdn.com%2finternalparasiteupdate-131106131939-phpapp02%2f95%2finternal-parasite-update-27-638.jpg%3fcb%3d1391164275&exph=479&expw=638&q=famacha+scoring&simid=608015862325903572&selectedIndex=17&ajaxhist=0
- Accuracy of FAMACHA© system for onfarm use by sheep and goat producers in the southeastern United States; J.M. Burke, R.M. Kaplan, J.E. Miller, T.H. Terrill, W.R. Getz, S. Mobini, E. Valencia; M.J. Williams, L.H. Williamson, and A.F. Vatta; Veterinary Parasitology [March 2007]. https://naldc.nal.usda.gov/download/10202/PDF
- Goat body scoring chart: https://www.bing.com/images/search?view=detailV2&ccid=ATkDdZRb&id=E7212E19E225A2279F3EDFC165A5DCD63EACE0C2&thid=OIP.ATkDdZRb-54RW3fFQkLzmQHaFj&mediaurl=https%3a%2f%2fi.pinimg.com%2foriginals%
- Open letter to sheep and goat producers about FAMACHA © program. Dr. Ray Kaplan. American Consortium for Small Ruminant Parasite Control. https://attra.ncat.org/attra-pub/viewhtml.php?id=215
- A Friendly and Encouraging Challenge to the Agricultural Extension Community: A low cost tool that can greatly influence management of internal parasites in small ruminants; Jim Morgan, Ph.D. [February 2005]. https://www.nap.edu/read/13087/chapter/2

- → Efficacy of copper oxide wire particles against gastrointestinal nematodes in sheep and goats; F. Soli, T.H. Terrill, S.A. Shaik, W.R. Getz, J.E. Miller, M. Vanguru, and J.M. Burke; Veterinary Parasitology [October 2009]. https://www.sciencedirect.com/science/article/pii/S0304401709006372

- → Managing Dewormer Resistance https://docs.wixstatic.com/ugd/6ef604_3981789ca4d34d74913834b1ea1b0b16.pdf

- → Decreasing barber pole larvae population on grass pastures: is liquid nitrogen fertilizer a viable alternative? Jean-Marie Luginbuhl, North Carolina State University [Timely Topic, April 2017] https://meatgoats.ces.ncsu.edu/2017/04/decreasing-barber-pole..

- → High quality forage helps to maintain resilience to GI parasites; by Dr. Ken Turner, USDA-ARS, El Reno, Oklahoma [Timely Topic, July 2014]. https://kerrcenter.com/wp-content/uploads/2014/04/hart_multispp...

- → Late Summer Parasite Management Strategies in Goats; by Dr. Ken Andries, Kentucky State University; https://www.wormx.info/andries2013#!

- → Improving parasite management with annual crops; by Richard Ehrhardt, Michigan State University https://www.wormx.info/annualcrops#!

- → Parasite control with multispecies grazing and rotational grazing; by Steve Hart; Langston University. https://kerrcenter.com/wp-content/uploads/2014/04/hart_multispp.

- → Strategies for coping with parasite larvae on pastures in the springtime in Ohio; by William Shulaw, Rory Lewandowski, Jeff McCutcheon, and Joyce Foster, Ohio State University Extension [2012]. https://ohioline.osu.edu/factsheet/VME-28

→ Effect of pelleting on efficacy of sericea lespedeza hay as a natural dewormer in goats; T.H. Terrill, J.A. Mosjidis, D.A. Moore, S.A. Shaik, J.E. Miller, J.M. Burke, J.P. Muir, and R. Wolfe; Veterinary Parasitology [February 2007] https://projects.sare.org/information-product/effect-of-pelleting-on-efficacy-of-sericea-lespedeza-hay-as-a-natural-dewormer-in-goats/

Websites of Interest to Goat Owners

→ Alpines International Club: www.alpinesinternationalclub.com

→ American Boer Goat Association: www.abga.org

→ American Dairy Goat Association: www.adga.org

→ American Goat Society: www.americangoatsociety.com

→ American Kiko Goat Association: www.kikogoats.com

→ American LaMancha Club: www.lamanchas.com

→ American Livestock Breeds Conservancy: www.albc-usa.org

→ American Nigerian Dwarf Dairy Association: www.andda.org

→ Canadian Goat Society: www.goats.ca

→ Canadian Meat Goat Association: www.canadianmeatgoat.com

→ Cashmere Goat Information: www.capcas.com

→ Crop Services: www.cropservicesintl.com

→ Guernsey Goat Breeders of America: http://guernseygoats.org

→ International Boer Goat Association: http://intlboergoat.org

→ International Fainting Goat Association: http://faintinggoat.com

→ International Kiko Goat Association: http://theikga.org

→ International Nubian Breeders Association: http://i-n-b-a.org

→ International Sable Breeders Association: http://sabledairygoats.com

→ Kinder Goat Breeders Association: http://kindergoatbreeders.com

→ Land of Havilah Herbals - Herbal remedy resource for wellness and emergency: www.landofhavilahfarm.com

- → Mohair Council of America: http://mohairusa.com
- → Molly's Herbals: Ohio@mollysherbals.com
- → National Pygmy Goat Association: http://npga-pygmy.com
- → National Saanen Breeders Association: http://nationalsaanenbreeders.com
- → National Toggenburg Club: http://nationaltoggclub.org
- → New Country Organics: Supplier of kelp and other goat feed related products newcountryorganics.com
- → Nigerian Dwarf Goat Association: http://ndga.org
- → Oberhasli Breeders of America: http://oberhasli.net
- → Oberhasli Goat Club: http://oberhasli.us
- → Ohio Earth Foods: https://ohioearthfood.com/pages/about-us
- → Pedigree Internationl LC (registers Genemasters): www.pedigreeinternational.com
- → Pygora Breeders Association: http://www.pba-pygora.com
- → United States Boer Goat Association: http://usbga.org
- → USDA Scrapie Information: www.aphis.usda.gov
- → DNA Testing for Goats:

    www.gtg.com.au

    www.biogeneticservices.com

    www.vgl.ucdavis.edu
- → Embryo Transfers: www.creeksideanimalclinic.com

General Goat Information Websites:

www.vetmed.UCdavis.edu

www.goatworld.com

www.goatworld.com/911

www.goatbiology.com

www.farminfo.org/livestock

www.sheepandgoat.com
www.goattalk.com

Magazines:

www.unitedcaprinenews.com
www.countrysidemag.com
www.dairygoatjournal.com
www.goatrancher.com
www.hobbyfarms.com
www.goattracksmagazine.com
www.motherearthnews.com
www.smallfarmgoat.com

Contributor Bios

Michelle Gorman - Integration Acres
Integration Acres, located in Albany, Ohio, has been focused on native foods since 1996. They help local growers and gatherers in the region generate income from natural resources while also preserving native plants from destruction. They use semi-wild cultivation methods to harvest pawpaw and other forest-farmed crops, using agricultural techniques in harmony with nature. Their facility hosts a certified dairy where they raise milking goats. The milk is value-added into a variety of cheese products within their facility including chevre and feta from their pasture raised goat's milk and aged cheeses using local cow dairy from Snowville Creamery.

Becky Rondy is the co-owner of Green Edge Organic Gardens, a year-round farm operation providing stable jobs for 12 people in economically distressed area of Appalachian Ohio. Funding from a 2012-2014 SARE PDP grant allowed Green Edge and Rural Action to partner and produce an extensive "Season Creation" curriculum, providing education around high-tunnel growing and farm management. Becky played a lead role in the design and delivery of the highly successful curriculum, which has educated over 200 farmers. Over 500 copies of a Green Edge Season Creation Manual were distributed to farmers throughout the Appalachian region. Her background, insights, and expertise have been an asset to the mentor farmers, helping them craft their presentations and work through the logistics of hosting large groups for on-farm workshops.

Leslie Schaller - Director of Programming, ACEnet, has directed multiple training and technical assistance programs which assist food and farm entrepreneurs through small business curriculum development, contract services and the formation of support networks of resource providers. As the ACEnet Director of Programs, Ms. Schaller secures public and private funding through grants and fees for services to support regional brand initiatives, targeted sector training, financial management support and capital access. She is the team leader coordinating all ACEnet programs and contracts.

In 2010, she co-authored, along with Michael Shuman and Brad Masi, *THE 25% SHIFT: The Benefits of Food Localization for Northeast Ohio & How to Realize Them -- The Northeast Ohio Local Food System Assessment and Plan*. In 2013, she was a contributor to the MIT study: Place-based Branding for Food Systems and Beyond: A study by the Central Appalachian Network Place-based Branding for Food Systems and Beyond: A study by the Central Appalachian Network which featured ACEnet regional brands.

Sasha Sigetic - Black Locust Livestock and Herbal. Sasha Sigetic grew up in the suburbs of Cleveland and attended college at Ohio University, majoring in ceramics. After moving to Austin, Texas and making art for a while, she got her Permaculture Design Certification. After that she moved back to Athens to help her family and to practice permaculture on a piece of land of her own. With the help of her partner and their land mate, she was able to start raising goats and has since moved to their new farm. She is the farmer and owner of Black Locust Livestock and Herbal, where she has a raw milk herd share with her Guernsey goats, Sasha forages her farm land for medicinal herbs and makes herbal tinctures to sell, as well as teaching ceramics at The Dairy Barn Art Center in town.

Celeste Taylor, owner of Black Sheep Farm. She is also the Goat Herd Manager at Integration Acres, an artisan dairy located near Athens, OH. Her skills include goat herding, sheep herding, natural goat and sheep health care, mechanical repairs and herd management. Celeste also utilizes livestock guardian dogs to protect her livestock. She is using rotational grazing to assist revitalizing the coal strip mined damaged land that makes up a large portion of her farm.

Abbe Turner, the owner of Lucky Penny Farm located in Garretsville, OH. Abbe currently raises Nubian, La Mancha and Alpine dairy goats and produces artisan cheese, goat's milk soap, and is known for the specialty caramel sauce, Cajeta. Lucky Penny Farms has two sister companies, Lucky Penny Creamery and Lucky Penny Candy, both operating out of the same building in Kent, OH. Turner worked to start the creamery by bringing together other local goat farmers, Slow Money Funds and local buyer support. The Creamery was added as an outlet for increased income through value added products.

Turner studied her craft at the University of Wisconsin-Madison, Pennsylvania Association of Sustainable Agriculture. She also honed her skills under dairy consultants Peter Dixon and Neville McNaughton. She is a Slow Food Terra Madre Delegate, and was selected to represent the Cleveland Slow Food Convivium in Turin, Italy in 2008 -- the same year she earned second place in the American Dairy Goat Association's Amateur Confection Contest.

NCR SARE

North Central Region SARE (NCR-SARE) is one of four regional offices that run the Sustainable Agriculture Research and Education (SARE) program, a nationwide grants and education program to advance sustainable innovation to American agriculture. NCR-SARE offers competitive grants and educational opportunities for producers, scientists, educators, institutions, organizations and others exploring sustainable agriculture in America's Midwest.

NCR-SARE has awarded more than $40 million worth of competitive grants to farmers and ranchers, researchers, educators, public and private institutions, nonprofit groups, and others exploring sustainable agriculture in 12 states. Project abstracts can be found by searching the SARE project database.

Rural Action

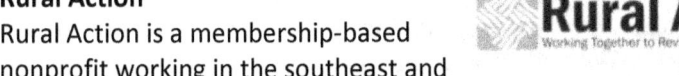

Rural Action is a membership-based nonprofit working in the southeast and central eastern counties of Appalachian Ohio since 1991. Their mission is to promote social, economic, and environmental justice by training, organizing and supporting communities. Their work is about developing the region's many assets in environmentally, socially, and economically sustainable ways. Rural Action focuses on sectors identified as important by their members: food and agriculture, forestry, zero waste and recycling, environmental education, and watershed restoration. Emerging work in social enterprise development, local tourism, and energy are more recent parts of the portfolio.

Rural Action has a network of over 600 members throughout Appalachian Ohio. The organization is governed by a Board from across southeast Ohio, with strong private sector involvement. The Board oversees a CEO who manages a staff of 20 full time equivalent staff and 26 AmeriCorps members through the Corporation for National and

Community Service. Rural Action members are involved as volunteers, participants, decision makers, and project designers.

AceNET

The Appalachian Center for Economic Networks (ACEnet) is a community-based economic development organization serving the 32 counties of Appalachia Ohio. The mission of ACEnet is to grow the regional economy by supporting entrepreneurs and strengthening economic sectors.

ACEnet staff accomplishes this mission by partnering with regional microenterprise and development practitioners to create a healthy local economy, allowing opportunity for all residents to start businesses, obtain quality jobs, and connect with other entrepreneurs for mutual benefit.

Today, ACEnet continues to provide incubation space and operate the Food Ventures Center. Recent efforts include helping business owners utilize new technology to promote their businesses. To assist local businesses in remaining competitive in the age of social media, ACEnet has held several well-attended workshops on utilizing social media platforms to engage consumers.

ABOUT THE AUTHORS:

Annie Warmke

Annie Warmke was born a city girl in Columbus Ohio but started her farming life with a Nubian goat on a small hobby farm in Baltimore Ohio during her high school years. Her career experiences include founding and running battered women's projects, women's funds, and organizing a variety of projects to provide leadership skills for women-owned businesses and non-profits. Annie has a degree in counseling from Ohio University, and has been a full-time goat herder for the past 13 years. She built Ohio's first Earthship, founded Blue Rock Station with her husband Jay Warmke, serves as a consultant to various groups, is vice-president of the USDA Farm Service Agency County Committee, and runs a goat college near Philo Ohio. Over the years she has won a variety of awards for her work, and in 2015 she co-founded Women Grow Ohio. Living at Blue Rock Station is her life-long dream, and being a grandmother has been her most important vocation.

Carie Starr

Carie Starr was born to hippie gardeners who raised her on her Cherokee grandmother's farm where, along with a strong sense of her native heritage, seeds of love for the earth and environment were planted. After earning a degree in Environmental Science from Zane State College she spent several years in the corporate world. Although successful in those jobs, the seeds of desire to connect to the earth were growing and blooming. In 2008 Carie, along with her husband Jarrod, took a huge leap into farming with the native symbol of power, blessings and abundance, the American Bison. Carie is now the full-time farmer of Cherokee Valley Bison Ranch, home to not only bison but heritage breed pigs, chickens and turkeys. In 2015 Carie co-founded Women Grow Ohio. In partnership with Women Grow Ohio and Rural Action, she created and hosted a workshop to educate aspiring farmers on the care and keeping of bison. When Carie isn't feeding pigs and chickens, watering livestock, planning education workshops, or pushing around hay bales she enjoys traveling with her husband and daughter Abby.

Women Grow Ohio
Sponsored by Rural Action
www.ruralaction.org
(740) 674-4300

_____ **Membership Application $25**

Women Grow Ohio is a volunteer-based group whose mission is to expand opportunities for Ohio's women farmers to network and grow as business professionals and food producers, by offering peer-to-peer mentoring and educational resources in agriculture, regional food networks and entrepreneurship.

Benefits of membership include:
- Access to a peer-to-peer mentoring program
- Networking with WGO members and partner organizations
- Visibility & Recognition
- Increased exposure as an WGO member
- WGO member organizations will be spotlighted on social media for innovative projects, programs or best practice in providing services.
- Link to your website or social media page
- Include the WGO member logo on your materials
- Learn about innovations, trends and opportunities through the WGO social media site
- Advertise related items for sale or rent including land, livestock, and produce
- Stay informed of events and news through WGO social media and email updates
- Receive updates about WGO programming and events
- Build your business image & increase exposure through a number of sponsorship & partnership opportunities

(Please include your name, business name, address, phone number, e-mail and website with your application fee of $25 for membership made payable to Rural Action WGO.)

Rural Action Kuhre Center for Rural Renewal, 9030 Hocking Hills Drive, The Plains, Ohio 45780